Stephan Wulfinghoff

Numerically Efficient Gradient Crystal Plasticity with a Grain Boundary Yield Criterion and Dislocation-based Work-Hardening

Schriftenreihe
Kontinuumsmechanik im Maschinenbau
Band 5

Karlsruher Institut für Technologie (KIT)
Institut für Technische Mechanik
Bereich Kontinuumsmechanik

Hrsg. Prof. Dr.-Ing.habil. Thomas Böhlke

Eine Übersicht über alle bisher in dieser Schriftenreihe erschienenen Bände finden Sie am Ende des Buchs.

Numerically Efficient Gradient Crystal Plasticity with a Grain Boundary Yield Criterion and Dislocation-based Work-Hardening

by
Stephan Wulfinghoff

KIT Scientific Publishing

Dissertation, Karlsruher Institut für Technologie (KIT)
Fakultät für Maschinenbau (MACH)
Tag der mündlichen Prüfung: 19.05.2014

Impressum

KIT Scientific Publishing

Karlsruher Institut für Technologie (KIT)
KIT Scientific Publishing
Straße am Forum 2
D-76131 Karlsruhe

KIT Scientific Publishing is a registered trademark of Karlsruhe Institute of Technology. Reprint using the book cover is not allowed.

www.ksp.kit.edu

Print on Demand 2014

ISSN 2192-693X
ISBN 978-3-7315-0245-6
DOI 10.5445/KSP/2192-693X

Numerically Efficient Gradient Crystal Plasticity with a Grain Boundary Yield Criterion and Dislocation-based Work-Hardening

Zur Erlangung des akademischen Grades

Doktor der Ingenieurwissenschaften

der Fakultät für Maschinenbau

Karlsruher Institut für Technologie (KIT)

genehmigte

Dissertation

von

Dipl.-Ing. Stephan Wulfinghoff

Tag der mündlichen Prüfung:	19.05.2014
Hauptreferent:	Prof. Dr.-Ing. Thomas Böhlke
Korreferent:	Prof. Dr.-Ing. Samuel Forest

Zusammenfassung

Die vorliegende Arbeit soll einen Beitrag zur Weiterentwicklung verschiedener Aspekte der gradientenerweiterten Plastizität liefern. Diese hat noch nicht den Status einer eindeutigen, anwendungsnahen und präzisen Theorie erlangt. Es werden daher verschiedene offene Fragen zu diesem Thema behandelt, wobei die effiziente numerische Implementierung besondere Berücksichtigung findet.

Nach einer kurzen Einführung des Versetzungsbegriffs und der beispielhaften Behandlung einer speziellen Gradientenplastizitätstheorie werden zwei vielversprechende, energetische Potentiale für die generalisierten Spannungstensoren untersucht, welche in vielen der oben genannten Theorien auftauchen. Anschließend wird ein Modellmaterial mit lokalisierten plastischen Scherbändern durch eine Gradiententheorie approximiert, um eine anschauliche Interpretation der zuvor genannten generalisierten Spannungen zu erhalten und ein verbessertes Verständnis verschiedener abstrakter Größen, wie etwa Mikroflächenkraftdichten, Mikrokraftbilanzen und inneren Längenparametern zu ermöglichen.

In der Folge wird eine thermodynamisch konsistente Gradientenplastizitätstheorie basierend auf einer skalaren, phänomenologischen plastischen Feldvariablen betrachtet, welche besonders für effiziente numerische Anwendungen geeignet ist. In diesem Zusammenhang werden energetische und dissipative generalisierte Spannungen sowie ein Korngrenzenfließmodell eingeführt. Darüber hinaus werden mögliche konzeptionelle Probleme der Theorie hinsichtlich bestimmter Instabilitäten aufgedeckt, die mit den Randbedingungen zusammenhängen, und es wird eine entsprechende Abhilfe für ein gradientenerweitertes Kristallplastizitätsmodell vorgeschlagen. Die Finite Elemente Implementierung dieses Modells wird anschließend diskutiert, wobei ein phänomenologisches lokales Voce-Verfestigungsgesetz und eine mikromorphe Formulierung eine einfache, robuste und numerisch effiziente Implementierung ermöglichen. Durch den Einsatz des Korngrenzen-

fließkriteriums können Größeneffekte erfolgreich reproduziert werden, welche in experimentellen Kupferdraht-Zugversuchen beobachtet wurden.
Um die physikalische Bedeutung des phänomenologischen Korngrenzenmodells weiter zu untersuchen, wird ein Vergleich mit dem Versetzungsaufstau-Modell von Eshelby et al. (1951) durchgeführt, und die Möglichkeit der Abbildung des Hall-Petch Effekts durch das Gradientenmodell wird diskutiert.
Anschließend wird das phänomenologische Voce-Verfestigungsmodell durch ein erweitertes Kaltverfestigungsmodell ersetzt. Das erweiterte Modell basiert auf der Gesamtversetzungsdichte und der mittleren Versetzungskrümmung, deren Evolution durch partielle Differentialgleichungen bestimmt wird, welche die krümmungsinduzierte Versetzungslinien-Produktion sowie Versetzungstransport berücksichtigen. Die Versetzungsevolutionsgleichungen stellen eine vereinfachte Version einer höherdimensionalen Theorie dar, welche im Detail aufgearbeitet wird.

Summary

This thesis is ought to be a contribution to the further development of several aspects of gradient plasticity, which has not yet achieved the status of a unique, applicable and generally accurate theory. Therefore, several open questions are addressed, where the efficient numerical implementation is particularly focused on. After a short introduction to dislocations and an exemplary gradient plasticity framework, the thesis inspects two promising constitutive energetical potentials for the generalized stress tensors arising in many theories. Subsequently, a model material with localized plastic slip bands is approximated by a gradient theory allowing for an illustrative interpretation of the aforementioned generalized stresses and a better general understanding of the meaning of several, usually abstract quantities like, e.g., microtractions, microforce balance equations and internal length scale parameters.

In order to consider a theory which is particularly suited for efficient numerical applications, a thermodynamically consistent gradient plasticity framework is discussed based on the introduction of a phenomenological scalar plastic field variable. In this context, energetical and dissipative generalized stresses as well as a grain boundary yield model are introduced. Moreover, possible conceptual difficulties of the framework concerning instabilities associated to the boundary conditions are revealed and a remedy is proposed for a gradient crystal plasticity model. The finite element implementation of this model is subsequently discussed using a phenomenological local Voce-hardening model and a micromorphic formulation allowing for a simple, robust and numerically efficient implementation. Owing to the grain boundary yield criterion, the size effect observed in tensile test experiments of copper micro-wires can successfully be reproduced.

In order to further investigate the physical significance of the phenomenological grain boundary model, a comparison to the pile-up model of Eshelby et al. (1951) is carried out and the potential of the

gradient model to reproduce the Hall-Petch effect is inspected.
An enhanced work hardening model is then introduced in order to replace the phenomenological isotropic Voce-hardening model. The enhanced hardening model is based on the total dislocation density and average curvature which evolve according to partial differential equations that take into account curvature-induced dislocation line-length production as well as dislocation transport. The dislocation evolution equations represent a simplified version of a higher-dimensional theory which is revised in detail.

Preface

The work presented in the thesis at hand has been developed at the Chair for Continuum Mechanics of Prof. Dr.-Ing. Thomas Böhlke at the Institute of Engineering Mechanics at the Karlsruhe Institute of Technology during the period from 2009 to 2014. The support of the German Research Foundation (DFG) in the project "Dislocation based Gradient Plasticity Theory" of the DFG Research Group 1650 "Dislocation based Plasticity" under Grant BO 1466/5-1 is gratefully acknowledged.
The English speaking reader shall forgive me my decision to write the following personal acknowledgements in German.
Als erstes möchte ich an dieser Stelle meinen Betreuer Prof. Dr.-Ing. Thomas Böhlke nennen, der mich schon im Studium unterstützt und meine Auffassung der Mechanik maßgeblich beeinflusst hat. Ich bin ihm zu tiefem Dank verpflichtet, denn er gab mir die Möglichkeit, meinen Lebensunterhalt mit meiner Passion zu verdienen. Vor allem möchte ich ihm herzlich für die große Freiheit und das hohe Maß an Vertrauen in meine Arbeit danken. Seine uneingeschänkt verständnis- und respektvolle sowie freundliche Art ermöglichten die geistige Unbefangenheit, die für eine produktive und kreative Arbeitsatmosphäre genauso unerlässlich sind, wie die bemerkenswerte Rückendeckung, auf die ich mich zu jeder Zeit bei ihm verlassen konnte.
Weiterhin möchte ich mich herzlich bei Prof. Dr.-Ing. Samuel Forest bedanken für die Übernahme der Zweitbetreuung und für einen unvergesslichen Monat am Centre des Matériaux in Evry (Je ne sais pas s'il y avait jamais un mois dans ma vie aussi productif et en même temps agréable que le Septembre 2013. Je vous en remercie cordialement!).
Genauso wie Herr Böhlke haben meine Kollegen zu einer tollen Atmosphäre am Institut beigetragen. Als erstes ist an dieser Stelle Eric Bayerschen zu nennen, der maßgeblich zur Durchführung der Simulationen in Kapitel 5 und zum Schreiben der zugehörigen Veröffentlichung beigetragen hat. Gleichzeitig war er genauso wie Vedran Glavas und Rumena Tsotsova aber auch immer ein toller Bürokollege,

so dass man sich am Arbeitsplatz immer gut aufgehoben gefühlt hat. Weiterhin danke ich Felix Fritzen für seine Unterstützung und seinen bemerkenswerten Code, welcher das FE-Rahmenwerk für die in dieser Arbeit vorgestellten Implementierungen lieferte. Zudem danke ich auch allen anderen Kollegen für die freundschaftliche Atmosphäre und die tolle Zeit!
Einen ganz besonderen Dank verdient Hannah Kudicke, die mir bei allen nicht-fachlichen Fragen rund um die Arbeit stets mit vollster Unterstützung zur Seite stand, die ein außergewöhnliches Verständnis für meine Arbeit aufbringt und jederzeit voll hinter mir steht.
Nicht zuletzt möchte ich mich bei meiner Mutter Ursula, ihrem Mann Jörg sowie meinem Bruder Aaron Quante herzlich dafür bedanken, dass ich jederzeit auf sie zählen kann. Besonders meine Mutter hat mit ihrem unermüdlichen und selbstlosen Einsatz die Basis für all das gelegt, was ich bis hierher erreichen durfte.

Karlsruhe, March 2014 Stephan Wulfinghoff

Contents

1 Introduction to Crystal Plasticity and Dislocations **15**
1.1 The Plastic Deformation of Single Crystals 15
1.1.1 Crystallographic Slip 15
1.1.2 Dislocations as Origin of the Plastic Deformation . 17
1.1.3 Regularization of Discrete Dislocations 20
1.2 Dislocations in Applied Material Science 21
1.2.1 Precipitation Hardening 21
1.2.2 Substitutional and Interstitial Atoms – Dynamic Strain Aging 22
1.2.3 Hall-Petch Effect 23
1.2.4 Single Crystals and Texture Strengthening 23
1.2.5 Work Hardening 24
1.3 Continuum Description of Dislocations in Crystal Plasticity 25
1.3.1 Orowan's Equation 25
1.3.2 Plastic Distorsion and Dislocation Density Tensor . 26
1.4 A Classical Continuum Crystal Plasticity Framework for Single Slip 29
1.4.1 Elastic Properties 29
1.4.2 Viscoplastic Formulation 29
1.4.3 Work Hardening Models 30
1.5 Gradient Crystal Plasticity Based on Scalar GND-Densities 32
1.5.1 Motivation 32
1.5.2 Different Representations of Geometrically Necessary Dislocations 32
1.5.3 A Gradient Plasticity Hardening Model 34

2 Application of Gradient Plasticity to a Laminate Material – Inspection of Energies **39**
2.1 Gradient Crystal Plasticity Based on the Dislocation Density Tensor 39
2.2 Shearing of a Periodic Laminate 44
2.2.1 Rank-One Defect Energy 46
2.2.2 Logarithmic Energy 50
2.3 Numerical Solution Strategy 58
2.3.1 Incremental Potential 58
2.3.2 Augmented Lagrange Multiplier and Penalty Method 59
2.3.3 Local Algorithms 60

3 Approximation of a Model Material With Localized Slip Bands by a Gradient Plasticity Theory **63**
3.1 Introduction 63
3.2 Slip Band Size Effects 65
3.2.1 Plastic Bending 65
3.2.2 Plastic Torsion 69
3.2.3 Single-Slip Torsion 72
3.2.4 Need for a new Continuum Theory 73
3.3 Generalized Energy Functions 74
3.3.1 Kinematical Assumptions 74
3.3.2 A Hierarchy of Energies 80
3.4 Dissipation Potential 92
3.5 A Gradient Plasticity Theory based on W_{12} 94
3.6 Polycrystals and Grain Boundaries 98
3.6.1 Beam Theory Solution 98
3.6.2 Continuum Theory Solution 101

4 Simplification of Gradient Plasticity based on a Scalar Plastic Field Variable **105**
4.1 Introduction 105

4.2 Power of the External and Internal Forces 110
4.3 Principle of Virtual Power and Balance Equations 111
4.4 Constitutive Assumptions 112
4.4.1 Energetic Assumptions and Dissipation Inequality 112
4.4.2 Rate-Independent Setting 114
4.4.3 Rate-Dependent Setting 117
4.5 Conceptual Difficulties in Plasticity including the Gradient of one Scalar Plastic Field Variable 118
4.5.1 One-Dimensional Example 118
4.5.2 A Stable One-Dimensional Formulation 121
4.6 Rate-Dependent Gradient Crystal Plasticity 124

5 Equivalent Plastic Strain Gradient Crystal Plasticity – Micromorphic Formulation and Implementation 127
5.1 Theoretical Framework for Grain Boundary Resistance . . 127
5.1.1 Motivation 127
5.1.2 Kinematical Assumptions 129
5.1.3 Principle of Virtual Power and Field Equations . . . 130
5.1.4 Constitutive Assumptions 131
5.2 Finite Element Implementation 134
5.2.1 Linearization of the Variational Form 134
5.2.2 Numerical Time Integration 135
5.2.3 Algorithmic Tangent 136
5.2.4 Discretization of the Grain Boundaries 137
5.3 Numerical Results 139
5.3.1 Simulation Setup and Boundary Conditions 139
5.3.2 Discretization Details 139
5.3.3 Model Parameters 141
5.3.4 Comparison of Simulation Results to Experimental Data 146
5.3.5 Discussion of Results 148
5.4 Enhanced Power-Law Subroutine 149
5.4.1 Convergence Improvement 149

5.4.2 Improvement of the Starting Solution 150
5.4.3 Numerical Examples 152

6 Gradient Plasticity Grain Boundary Models vs. a Classical Pile-Up Theory 157

6.1 Introduction 157
6.2 Size Effects provoked by Energetical and Dissipative Grain Boundary Models within Gradient Plasticity 158
6.2.1 Energetical Interfaces 159
6.2.2 Dissipative Interfaces 160
6.3 Review of the Hall-Petch Model of Eshelby et al. (1951) . . 160
6.4 Comparison of the Eshelby Model with Gradient Plasticity Grain Boundary Models 163
6.4.1 Review of Grain Boundary Stresses in Gradient Plasticity 163
6.4.2 Comparison with the Eshelby Model 164
6.5 A Simple Gradient Theory 166
6.6 Numerical Simulations 168
6.6.1 Simulation Set-Up 168
6.6.2 Interpretation 169

7 Revision of a Continuum Dislocation Dynamics Theory at Small Strains 173

7.1 Introduction 174
7.2 Kinematics of a Discrete Dislocation 175
7.3 Review of the Continuity Equation 177
7.4 An Orientation-Dependent Dislocation Density 178
7.5 Dislocation Transport 182
7.6 Line Length Production 184
7.7 Rotational Density Transport 185
7.8 Evolution of the Segment Density 188
7.9 An Analytical Solution 190
7.10 Averaged Equations 191

8 A Simplified Thermodynamic Continuum Dislocation Dynamics Framework – Theoretical Aspects and Finite Element Simulation 193

8.1 Introduction . . . 193
8.2 Kinematical Framework . . . 195
8.2.1 Distorsion and Dislocation Measures . . . 195
8.2.2 Kinematical Equations . . . 196
8.3 Principle of Virtual Power . . . 196
8.4 Energy Density and Dissipation . . . 198
8.5 Yield Criteria and Flow Rule . . . 200
8.6 Numerical Example . . . 201

9 Summary and Outlook 207

A Appendix of Chapter 2 215

A.1 Vanishing External Power of Micro Tractions . . . 215
A.2 Power Balance for Rank-One Energy . . . 215
A.3 Sensitivity Analysis . . . 216

B Appendix of Chapter 3 219

B.1 Displacement Consistency . . . 219
B.2 Verification of W_{21} . . . 219
B.3 Slip Consistency . . . 220
B.4 Verification of W_{12} . . . 220

C Appendix of Chapter 6 223

C.1 Energetical Interfaces . . . 223
C.2 Dissipative Interfaces . . . 225
C.3 Interpretation of the Material Parameters . . . 226

D Appendix of Chapter 7 229

Bibliography 240

Introduction

Motivation

Plasticity theories are nowadays widely used in research and industry. They cover a broad field of applications ranging from crash and deep drawing simulations to the modeling and optimization of microstructured materials. However, classical continuum plasticity theories without internal length scale usually fail to predict the mechanical behavior of micro-components if their dimensions fall below a critical size (typically around $10\mu m$). In addition, these models are not able to account for size effects when they are applied on the microscale to predict the effective behavior of micro-heterogeneous materials. For example, they cannot reproduce the influence of the precipitate or grain size (i.e. the Hall-Petch effect). However, these size effects are technologically highly relevant.
These drawbacks motivated the mechanical community to enhance classical plasticity theories by introducing internal length scales into the models. In many cases, this is realized through the consideration of plastic strain gradients or familiar quantities in the constitutive equations. The principal motivation of this kinematical extension is the close relation of plastic strain gradients to geometrically necessary dislocations (GNDs). These represent a special continuum description of the underlying dislocation structure, described by different effective Burgers vector measures. For example, GNDs may be given in terms of the dislocation density tensor, which is the curl of the plastic distorsion. Alternatively, closely related scalar edge and screw densities are often introduced for the individual slip systems of a single crystal. Each representation leads to an improved consideration of the dislocation

microstructure in the respective model.
Many gradient plasticity theories are formulated in a thermodynamic fashion, e.g., by adding the dislocation density tensor to the list of arguments of the stored energy function. As a result, the mechanical response of the model indeed becomes size-dependent, which – in principle – allows for a large range of new applications.
However, due to the need for abstract quantities like generalized stresses or microtractions, gradient plasticity theories are often difficult to understand for researchers unfamiliar with the field. In addition, many open questions exist concerning the following aspects.

- Often, simplistic quadratic energy density functions are applied. The optimal form of the energy is still subject of current research.
- Additional partially highly complex boundary conditions are required. These have significant influence on the solution. However, oversimplified conditions are usually applied due to the lack of more realistic models.
- Gradient plasticity facilitates the formulation of grain boundary models. However, this technically highly relevant aspect needs significantly more investigation, since available models are usually not sufficiently accurate.
- The material parameters involved often lack a satisfying physical interpretation. Instead, they are often simply introduced as "internal length scale".
- In gradient plasticity, work hardening is often modeled in terms of so-called statistically stored dislocation densities. Their effective transport is usually neglected, although the admissibility of this omission is highly questionable on the microscale.

Gradient plasticity is an active field of current research and did not yet achieve the status of a quantitatively correct and practical theory. In many cases, the results do not even match qualitatively with experiments. For example, many models yield a size-dependent overall hardening instead of a size-dependence of the yield strength, which is

more frequently observed in experiments.
This lack of accuracy still contrasts with the significantly increased computational effort of gradient models compared to classical theories without internal length scale. For example, the global systems of linear equations arising in finite element simulations of gradient plasticity models are mostly significantly larger than those of comparable classical models. Therefore, there are barely any three-dimensional applications. Moreover, the additional nodal unknowns usually represent plastic variables which, depending on the model, require a global active set search strategy or comparable nontrivial algorithms for the determination of the plastically active zone.
This work is ought to be a contribution to the resolution of various of the aforementioned issues.

- The work investigates more promising energy density functions than the quadratic approach. Amongst other things, all material parameters of the functions allow for a physical interpretation.
- In order to improve the understanding of various rather abstract quantities like generalized stresses or the internal length scale parameter, a model material is approximated by a gradient theory. For this particular material, gradient plasticity features which are usually abstract, take an illustrative character and can be understood intuitively.
- In order to reduce the large computational effort in three-dimensional applications, a simplified model is proposed. It is based on the gradient of a scalar plastic field variable and includes thermodynamically consistent and numerically efficient grain boundary formulations.
- Compared to other theories (e.g. based on the full dislocation density tensor), a significant reduction of nodal degrees of freedom is obtained in a finite element simulation of the model. In addition, the theory allows for a robust implementation and large time steps. It is shown to facilitate the simulation of three-dimensional

problems, involving several hundred grains, with a comparatively low computational effort. Although more complex models are more precise, the theory preserves many important features of other gradient plasticity frameworks.

- Depending on the application and the parameters of the simplified model, the size-effect concerns either the overall hardening, the overall yield strength, or both. The material parameters are successfully fitted to experimental data of micro tensile-tests, illustrating the capacity of the model to reproduce the experimentally observed size-effect. In addition, the potential to capture the Hall-Petch effect is extensively discussed.
- An enhanced work hardening model is introduced based on dislocation densities, the evolution of which is governed by partial differential equations. Amongst other things, these equations do not neglect dislocation transport, in contrast to many other approaches. The evolution equations of the dislocation densities are investigated thoroughly.

State of the Art

The discovery of dislocations was a milestone in the process of understanding plastic deformations in metals (Orowan, 1934; Polanyi, 1934; Taylor, 1934). Henceforth, many experimentally observed phenomena could be explained physically by a rapidly emerging dislocation theory. Examples are given by the particular plastic deformation of single crystals which is usually characterized by plastic shear in specific crystallographic slip systems. Moreover, the discrepancy between the theoretical and the measured critical resolved shear stress as well as the observed work hardening could be explained (Schmid and Boas, 1935; Taylor, 1938; Kröner, 1960). See Hirth and Lothe (1982) as well as Hull and Bacon (1984) for an introduction to dislocation theory.
Continuum mechanical formulations have been established describing

the evolution of larger amounts of dislocations in terms of scalar densities. These are interpreted as total line length per unit volume (Gillis and Gilman, 1965; Essmann and Rapp, 1973; Essmann and Mughrabi, 1979; Estrin, 1996; Arsenlis and Parks, 2002; Hochrainer et al., 2014). A further important continuum mechanical quantity is the dislocation density tensor. This effective Burgers vector variable represents an illustrative measure of plastic incompatibility (Kondo, 1952; Nye, 1953; Bilby et al., 1955; Kröner, 1958).

A geometrically linear, macroscopic continuum crystal plasticity theory was proposed by Hill (1966). This framework as well as its geometrically nonlinear generalization are nowadays widely applied (Teodosiu, 1970; Rice, 1971; Mandel, 1971). An important ingredient of the framework are phenomenological work hardening models, accounting for the effect of dislocation multiplication and entanglement on the overall material strength (e.g. Taylor, 1938; Hill, 1966; Kocks, 1970; Chang and Asaro, 1981; Franciosi and Zaoui, 1982; Estrin, 1996; Kocks and Mecking, 2003). Numerical Algorithms for local single crystal plasticity were discussed, e.g., by Needleman et al. (1985), Cuitiño and Ortiz (1992), Steinmann and Stein (1996), Ortiz and Stainier (1999) as well as Miehe and Schröder (2001).

An early strain gradient plasticity theory allowing for the modeling of size effects in shear bands was formulated by Aifantis (1987). Further gradient theories by Fleck and Hutchinson (1993), Steinmann (1996) as well as Nix and Gao (1998) exploit the concept of geometrically necessary dislocations developed by Ashby (1970). Henceforth, a large number of micro-experiments has been performed covering many loading cases like tension, compression, bending, torsion and micro-indentation (e.g. Fleck et al., 1994; Stölken and Evans, 1998; Dimiduk et al., 2005; Xiang and Vlassak, 2006; Gruber et al., 2008; Yang et al., 2012). The principal outcome of these tests is the confirmation of the size-dependence of plastically inhomogeneous deformations on the microscale, summarized by the slogan "smaller is stronger".

Simultaneously, additional gradient crystal plasticity theories emerged, most of them being formulated thermodynamically based on an extension of the stored energy function. Ortiz and Repetto (1999) propose a rank-one energy, leading to a size-dependent overall yield stress as shown by Ohno and Okumura (2007) using a slightly different energy. Rank-one potentials have been applied in various finite element simulations and require a special numerical treatment since they are non-smooth (Kametani et al., 2012; Hurtado and Ortiz, 2012, 2013; Wulfinghoff et al., 2014).

Berdichevsky (2006a) proposes a logarithmic energy with similar properties, which additionally accounts for dislocation saturation effects (see also Berdichevsky, 2006b). This potential has been investigated in a series of analytical and numerical solutions, e.g., by Berdichevsky and Le (2007), Le and Sembiring (2008b), Kochmann and Le (2008) as well as Kochmann and Le (2009b). It was further compared to discrete dislocation dynamics simulations by Le and Sembiring (2008a) and extended to a deformation twinning theory by Kochmann and Le (2009a).

Mesarovic et al. (2010) discuss a three-dimensional gradient crystal plasticity theory including a quadratic energy. This theory has been derived from the coarsening of highly correlated dislocation arrangements in discrete slip planes (see also Mesarovic, 2005; Yassar et al., 2007; Baskaran et al., 2010; Mesarovic, 2010).

The different energy functions have been compared by Forest and Guéninchault (2013) as well as Wulfinghoff et al. (2014). The authors propose another logarithmic function motivated by the statistical theory of Groma et al. (2003).

Gurtin (2000) formulated a large strain gradient plasticity framework for single crystals with special emphasis on the exploitation of the principle of virtual power. It has been further developed by Gurtin (2002) and specified for small deformations by Cermelli and Gurtin (2002). Moreover, Gurtin and Needleman (2005) formulated a closely related grain boundary model based on the Burgers vector flow (see also Gurtin, 2006, 2008b). The aforementioned rank-one energy of Ohno and Okumura (2007) has been integrated into the framework by Gurtin (2008a). Besides energetical generalized stresses, the theory allows for dissipative stresses as well (e.g. Anand et al., 2005; Gurtin et al., 2007). The concept of generalized dissipative stresses is critically discussed by Hutchinson (2012) who noted that dissipative "higher order stress quantities can change discontinuously for bodies subject to arbitrarily small load changes".

A closely related, geometrically linear framework for gradient-extended dissipative solids has been proposed by Miehe (2011). Here, the principle of virtual power derives from a global incremental potential (see also Hurtado and Ortiz, 2013). The framework is a further development of local variational concepts (e.g. Ortiz and Stainier, 1999; Miehe, 2002) and has been applied to an isotropic plasticity model by Miehe et al. (2013). In a series of papers, the theory has been generalized to large deformations by Miehe (2014) and applied to gradient crystal plasticity by Miehe et al. (2014a). The contribution comprises a new finite element formulation which has been used to solve several three-dimensional problems, partially involving multiple grains.

Forest (2009) proposes a micromorphic theory based on the introduction of an additional field variable, which is interpreted as the micromorphic counterpart of an internal or state variable. Depending on the choice of the variables and their coupling, the theory is able to reconcile several classes of generalized theories, e.g., gradient enhanced plasticity and damage models, Eringen's and Mindlin's micromorphic theory (Eringen and Suhubi, 1964; Mindlin, 1964) as well as implicit gradient models (e.g. Engelen et al., 2003; Peerlings et al., 2012). The theory has been applied to crystal plasticity and damage, e.g., by Cordero et al. (2010), Aslan et al. (2011) and Cordero et al. (2013). It facilitates the preservation of the local character of the flow rule, i.e., an associated variational formulation is not required. This property has been exploited numerically, e.g., by Anand et al. (2012), Cordero et al. (2012) and Wulfinghoff et al. (2013a).

Geometrically linear thermodynamical gradient plasticity theories for polycrystals have been developed, e.g., by Menzel and Steinmann (2000), Gurtin (2004) as well as Gudmundson (2004) (see also Fredriksson and Gudmundson (2005)). The polycrystal theory of Fleck and Willis (2009b,a) has been implemented by Nielsen and Niordson (2013). Geometrically nonlinear polycrystal frameworks are discussed, e.g., by Liebe et al. (2003) as well as Miehe et al. (2014b), who propose a gradient extended version of the logarithmic strain theory by Miehe et al. (2002). In microplasticity, grain boundaries play an important role, since they have a strong influence on the overall plastic response. Therefore, the possibility to formulate respective interface models is a crucial feature of gradient plasticity. Several works focus particularly on grain boundary models (e.g. Aifantis and Willis, 2005; Aifantis et al., 2006; Fredriksson and Gudmundson, 2007). The theory of Dahlberg et al. (2013) accounts for interfaces that slide and separate. A numerical treatment of a grain boundary model is discussed by Wulfinghoff et al. (2013a).

In contrast to thermodynamical frameworks, various gradient plasticity theories do not explicitly introduce generalized stresses conjugate to the gradients of plastic strain measures. Examples are given by the contributions of Acharya and Bassani (2000), Evers et al. (2002), Cheong et al. (2005) and Becker (2006). Ertürk et al. (2009) discussed a unification of Gurtin's thermodynamical theory and the physically motivated model formulation of Evers et al. (2004a,b) (see also Bayley et al., 2006; Geers et al., 2007). Additional comparisons were performed by Bargmann et al. (2010) as well as Svendsen and Bargmann (2010), who showed that many of the aforementioned models can be unified in a thermodynamic rate variational formulation.

Apart from the application to size-effects, gradient extended models are often applied to regularize localization phenomena, e.g., due to softening (e.g. Liebe, 2004; Anand et al., 2012; Mazière and Forest, 2013) and damage (e.g. Geers et al., 1998; Liebe et al., 2001; Miehe et al., 2010a,b; Aslan and Forest, 2011).

Usually, numerical implementations of gradient plasticity define plastic strain measures as independent variables, i.e., they appear as additional

nodal degrees of freedom. Early numerical contributions of this kind were made by de Borst and Mühlhaus (1992) as well as Liebe and Steinmann (2001). The latter propose a variational formulation of the flow rule, leading to an active set search procedure on the nodal level. Becker (2006) compared this strategy to a projection method which maps the integration point values of the plastic slips to the nodes. A similar projection procedure has been applied by Han et al. (2007).
Kuroda and Tvergaard (2008) introduce scalar densities of geometrically necessary dislocations as additional nodal degrees of freedom (see also Klusemann et al., 2012; Kuroda, 2013). Moreover, Reddy et al. (2012) discuss a primal-dual formulation where the displacements, the plastic slips and the backstress are treated as global unknowns (see also Wieners and Wohlmuth, 2011). This strategy facilitates a local evaluation of the radial-return algorithm, similar to the micromorphic approach of Forest (2009). For a discontinuous Galerkin approach to gradient plasticity, see Djoko et al. (2007a,b).
The theory of Gurtin (2002) was implemented by Bittencourt et al. (2003) and compared to Discrete Dislocation Dynamics (DDD) simulations of a model composite subject to simple shear (see, e.g., Cleveringa et al., 1997). Closely related comparisons have been carried out, e.g., by Shu et al. (2001), Aifantis et al. (2009), Aifantis et al. (2012) as well as Bardella et al. (2013), who used DDD-simulations to discuss the effect of geometrically necessary dislocations on latent hardening.
Yefimov et al. (2004) used DDD-results to benchmark the statistical dislocation theory of Groma et al. (2003). Groma's work has been derived from the dynamics of individual dislocations. It relates the internal length scale of the backstress to the inverse of the square root of the total dislocation density, instead of treating it as a material constant. The model accounts for dislocation sources and dislocation transport in a continuum mechanical sense.
Hochrainer (2006) developed a three-dimensional generalization of Groma's kinematical dislocation density evolution equations (see also Hochrainer and Zaiser, 2005; Hochrainer et al., 2007; Zaiser et al., 2007).

Hochrainer's basic field variables are given by a scalar dislocation density and average dislocation curvature. These are defined in an extended space which accounts for the dislocation orientation in addition to the three-dimensional Euclidean space and the time. A simplified theory without extra dimensions has been proposed by Hochrainer et al. (2010) (see also Hochrainer et al., 2014). Numerical simulations of both approaches have been carried out by Sandfeld et al. (2010), Sandfeld (2010), Sandfeld et al. (2011), Wulfinghoff et al. (2011) as well as Wulfinghoff and Böhlke (2012b), amongst others.

Outline

The thesis starts with a short introduction to crystal plasticity and dislocations in Chap. 1. The geometrical description as well as the stress fields of edge and screw dislocations are summarized. Fundamental examples concerning the application of dislocation theories in applied material science are discussed. Subsequently, continuum mechanical descriptions of dislocations are introduced. Moreover, a classical continuum mechanical crystal plasticity framework including a work-hardening model is shortly reviewed. In addition, some basic gradient plasticity model features are introduced by means of an example which illustrates some differences between classical and gradient-enhanced approaches.

Chap. 2 investigates two promising defect energy candidates for gradient crystal plasticity applications. First, a rank-one defect energy is considered, allowing for a size-effect on the overall yield strength of micro-heterogeneous materials. As a second candidate, a logarithmic defect energy is proposed, which is motivated by the work of Groma et al. (2003). The characteristics of both energies are investigated by applying them to a laminate microstructure. A new regularization technique for the numerical treatment of the rank-one potential is presented based on an incremental potential involving Lagrange multipliers. The effect of the two energies on the macroscopic size-dependent stress-

strain curve as well as the arising pile-up forms are investigated. Under cyclic loading, the model is shown to provide a continuum formulation of Asaro's type III kinematic hardening (Asaro, 1975).
The aim of Chap. 3 is to improve the understanding of several abstract quantities introduced in Chaps. 1 and 2, like generalized stresses and the internal length scale. Therefore, a special single crystal model is approximated by a gradient plasticity theory in a single slip situation. The particular feature of the single crystal model is the concentration of the plastic slip into ideally localized and equidistant slip bands. Hence, the regions in between the slip bands deform purely elastically. Exemplary bending and torsion examples illustrate that continuum crystal plasticity theories without internal length scale are unable to properly describe the material behavior, if the characteristic deformation wave lengths approach the slip band spacing. Subsequently, it is shown that an improved continuum mechanical approximation of the material is obtained by enhancing the stored energy function by a defect energy term. For this specific model, the generalized stresses, tractions, boundary conditions, microforce balance equation as well as the internal length scale have an illustrative and intuitive character in certain situations. The introduction of a dissipation potential and the presentation of an illustrative example close the discussion.
In Chap. 4, the frameworks introduced in the first part of the work are simplified in order to optimize the theory for three-dimensional numerical applications of several grains, including grain boundary models. Therefore, the number of kinematically independent global plastic field variables is reduced to only one phenomenological scalar field. The grain boundaries are explicitly accounted for in the principle of virtual power, yielding an additional grain boundary microforce balance equation, which supplements the linear momentum balance and the microforce balance equation in the bulk. The introduction of bulk and grain boundary energy density functions as well as thermodynamically consistent flow rules close the set of equations. In this context, rate-dependent and rate-independent settings are discussed.

Furthermore, it is shown that the reduction to only one scalar plastic field variable leads to conceptual difficulties concerning the direction of plastic flow. By means of an example, it is demonstrated that the solution is generally not stable with respect to the boundary conditions, i.e., arbitrarily small changes of the boundary conditions may lead to finite changes of the solution. An alternative, stable approach is discussed and applied to a gradient-extended, visco-plastic single crystal model.

In Chap. 5, a micromorphic version of the crystal plasticity model developed in Chap. 4 is presented. The finite element implementation is discussed in detail. Emphasis is put on the enhancement of the power law material subroutine. The associated implicit Euler scheme is optimized based on an improved starting value for the Newton scheme. Three-dimensional simulations illustrate that the proposed algorithm facilitates significantly larger time steps compared to the standard Newton scheme. The numerical model is fitted to experimental data of polycrystalline copper micro-tensile tests. Therefore, a quadratic defect energy and an isotropic Voce-hardening model are applied. The size dependent yield strength is reproduced notably well owing to the grain boundary model introduced in Chap. 4.

This interface model is compared to a pile-up model in the spirit of Eshelby et al. (1951) in Chap 6. A relation between the relevant grain boundary stresses of both theories is established. This relation illustrates the incompatibility of the approaches: the gradient model can in general not be expected to yield the typical Hall-Petch relation (the yield stress increase scales inversely with the square root of the grain diameter). The findings are verified by three-dimensional FEM-simulations of several hundred grains.

Chap. 7 revises the kinematical continuum dislocation framework of Hochrainer (2006) based on the dislocation density and average dislocation curvature, which are introduced for each slip system. Both fields are defined in the Euclidean space extended by the dislocation orientation, which is introduced as an extra-dimension. The different terms of the

evolution equations are motivated from the kinematics of individual dislocations. Finally, a simplified version of the theory without extra dimension is discussed (Hochrainer et al., 2010).
In Chap. 8, the simplified version is coupled to the equivalent plastic strain gradient model presented in Chap. 5. The phenomenological isotropic Voce-hardening approach is therefore replaced by a dislocation density based work-hardening model. More precisely, the critical resolved shear stress of each slip system is assumed to be a function of the total dislocation densities, the evolution of which is governed by the partial differential equations introduced in Chap. 7. In addition, the introduction of further generalized stresses is discussed. These are assumed conjugate to the dislocation densities and the plastic slips. As a result, additional generalized force balance equations are obtained which enter the flow rule in form of a backstress term. In particular situations, the backstress comes out to be a line-tension induced curvature effect.

Notation. A direct tensor notation is preferred throughout the text. Vectors and 2nd-order tensors are denoted by bold letters, e. g. $\boldsymbol{a}$ or $\boldsymbol{A}$. The symmetric part of a 2nd-order tensor $\boldsymbol{A}$ is designated by $\mathrm{sym}(\boldsymbol{A})$. A linear mapping of 2nd-order tensors by a 4th-order tensor is written as $\boldsymbol{A} = \mathbb{C}[\boldsymbol{B}]$. The scalar product of 2nd-order tensors is denoted, e. g., by $\boldsymbol{A} \cdot \boldsymbol{B} = \mathrm{tr}(\boldsymbol{A}^{\mathsf{T}}\boldsymbol{B})$. The composition of two 2nd-order tensors is formulated by $\boldsymbol{AB}$ and the dyadic product by $\boldsymbol{A} \otimes \boldsymbol{B}$. Matrices are labeled by a hat, e. g. $\hat{\varepsilon}$.

Chapter 1

Introduction to Crystal Plasticity and Dislocations

This chapter serves as an introduction to crystal plasticity and dislocations. The presentation is kept as short as possible since most of the content is a revision of well-established theories and does not contain any new findings. In particular, the text concentrates on aspects which are relevant for the subsequent chapters. For a more detailed and comprehensive introduction to dislocations the reader is referred to Hull and Bacon (1984). Hirth and Lothe (1982) give a broad overview on the theory of dislocations. In the context of applied material science, Bargel and Schulze (2008) provide a basic introduction. Standard references on continuum crystal plasticity models are, e.g., Mandel (1971) and Rice (1971). The content has partially been taken from Wulfinghoff and Böhlke (2010).

1.1 The Plastic Deformation of Single Crystals

1.1.1 Crystallographic Slip

Usually, everyday metallic components exhibit a polycrystalline microstructure. This means they are composed of a large number of individual crystals, which form regions with highly ordered atomic structure. The size of these so-called grains can vary within a wide range (from less than one up to several hundred microns, depending on the application). The grains are delimited by grain boundaries. In many cases, the macroscopic mechanical properties of polycrystalline materials are nearly isotropic (direction-independent), since the crystallographic orientation of the individual grains does not show any preferred direction. However, the individual crystals are anisotropic due to their crystallographic atomic structure. In particular, the basic plastic deformation mechanism is shear in certain lattice directions.

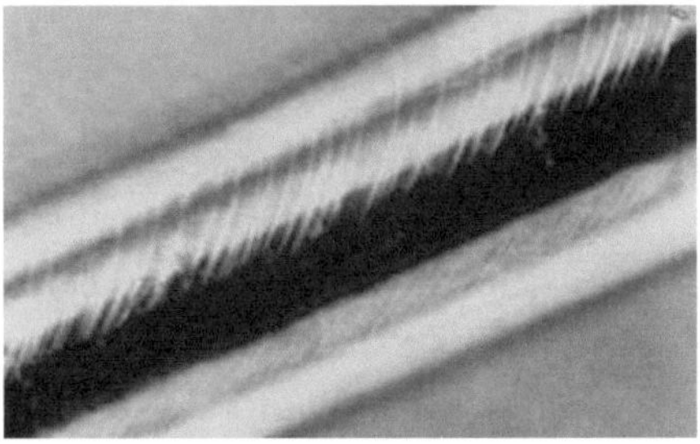

Figure 1.1: Slip traces on an tungsten-specimen (taken from Schob, 1972).

For example, Fig. 1.1 shows a deformed tungsten single crystal tensile specimen. The slip traces illustrate the plastic shear deformation. Geometrically, the deformation can be characterized by the slip direction $\boldsymbol{d}$, the slip plane normal $\boldsymbol{n}$ and the amount of plastic slip γ. Together, these quantities form a slip system. They are schematically illustrated in Fig. 1.2, which shows a homogeneously sheared piece of material. In order to attach an orthonormal coordinate system to the considered slip system one may define a third vector by $\boldsymbol{l} = \boldsymbol{n} \times \boldsymbol{d}$, or alternatively $\boldsymbol{h} = \boldsymbol{d} \times \boldsymbol{n} = -\boldsymbol{l}$ (see Fig. 1.2).
Only one active slip system is visible in Figs. 1.1 and 1.2, respectively. This particular case is referred to as single-slip. In general, several slip systems are present in a crystal and can simultaneously be active (see Chap. 1.5). If the crystal orientation of a single crystal tensile specimen is properly chosen, the activated slip system usually forms an angle of approximately $45°$ with the loading direction, indicating the direction of the maximum shear stress.

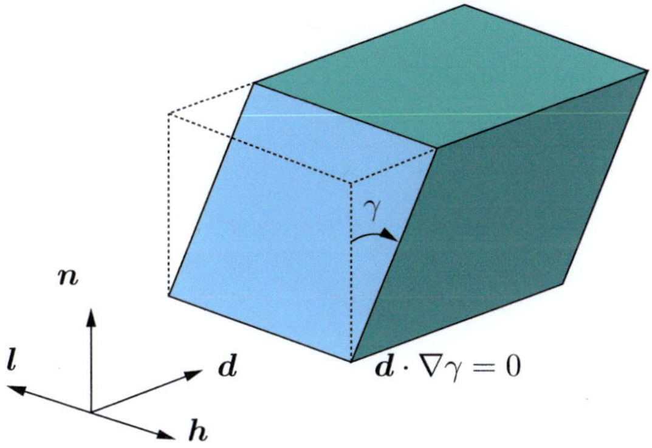

Figure 1.2: Illustration of the slip direction $\boldsymbol{d}$, the slip plane normal $\boldsymbol{n}$ and the plastic slip γ (for small deformations).

1.1.2 Dislocations as Origin of the Plastic Deformation

In real crystals, the plastic shear deformation process is not as smooth as indicated in Fig. 1.2. Instead, it is the result of the collective motion of a large number of discrete, line-like crystallographic defects, the so-called dislocations. Therefore, Fig. 1.2 shows an idealization of the actual plastic deformation. Fig. 1.3 shows a schematic example of a dislocation. The motion of the dislocation on the glide plane with normal $\boldsymbol{n}$ causes a local change of the atomic neighborship relations. This is the principal origin of the macroscopically observed plastic deformation.
The dislocation in Fig. 1.3 has entered the crystal from the surface, leaving a step. The plastic slip represents a displacement jump, i.e., it is localized and can be interpreted as the trace of the dislocation (indicated by the dashed line in Fig. 1.3).

The presence of the dislocation implies an elastic deformation of the crystal lattice (Fig. 1.3) and a stress field around the dislocation. For the simple geometry in Fig. 1.4 (left), the stress field can be computed analytically in regions which are sufficiently far from the dislocation core (Volterra, 1907; Taylor, 1934). For this purpose, the theory of

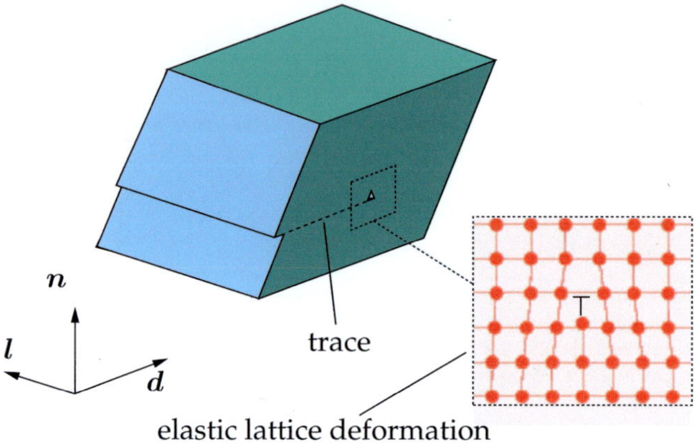

Figure 1.3: Motion of an edge dislocation through the crystal lattice (schematically). The trace of the dislocation represents a localized plastic slip.

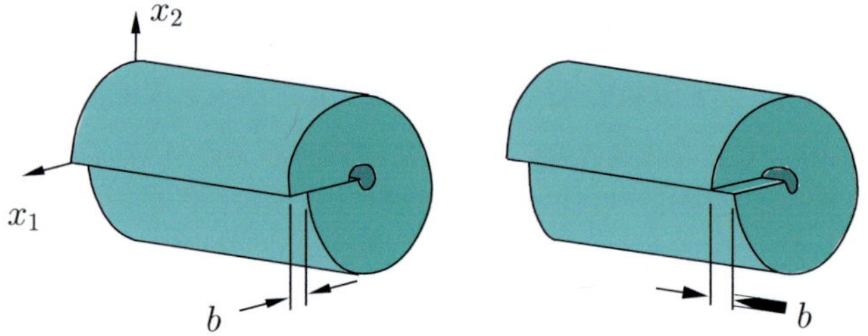

Figure 1.4: Volterra edge dislocation (left) and screw dislocation (right).

isotropic linear elasticity is applied. The stress field of the edge dislocation (Fig. 1.4, left) is given by

$$\sigma_{11} = D_{\mathrm{s}} x_2 \frac{3x_1^2 + x_2^2}{r^4}, \quad \sigma_{12} = -D_{\mathrm{s}} x_1 \frac{x_1^2 - x_2^2}{r^4}, \quad \sigma_{22} = -D_{\mathrm{s}} x_2 \frac{x_1^2 - x_2^2}{r^4}, \quad (1.1)$$

where $D_{\mathrm{s}} = bG/(2\pi(1-\upsilon))$ with Poisson's ratio υ, the shear modulus G and the absolute value b of the Burgers vector (Fig. 1.4), which will subsequently be defined more rigorously. The radial distance from the dislocation core is denoted by r.

In addition, Fig. 1.4 (right) schematically shows the second basic dislocation type: the screw dislocation. This kind of dislocation leads to shear stresses $\tau(r)$ which are tangentially aligned to circles around the dislocation core and which lie in planes orthogonal to the line direction. They are given by

$$\tau(r) = \frac{bG}{2\pi r}. \tag{1.2}$$

In general, dislocations in real materials are curved, i.e., they are compositions of the aforementioned edge and screw prototypes. The elastic dislocation energy per unit length W_1 can be computed from the stress fields by integration

$$W_1 = \alpha G b^2, \tag{1.3}$$

where α depends on the geometry of the considered system and on whether the dislocation is of screw, edge or mixed type.

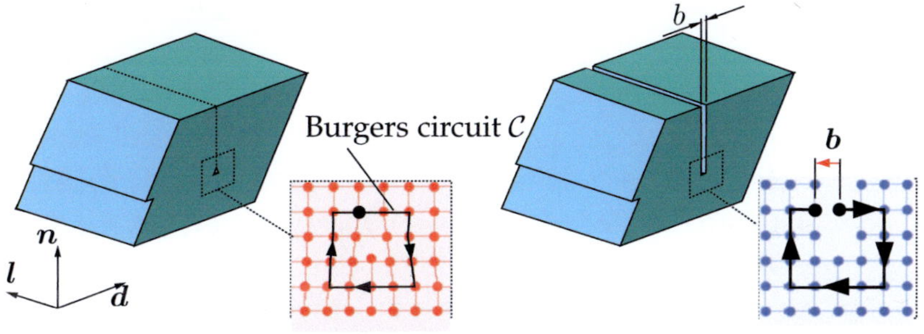

Figure 1.5: Relaxation of the crystal lattice and Burgers circuit (neglecting surface energy effects).

Usually, the velocity of the screw portion of an expanding dislocation loop succeeds the velocity of the edge portion, since the motion of the edge part implies the production of additional screw line length. However, the edge dislocation is energetically more favorable leading to a larger driving force of the screw component motion.

The stored elastic energy around the dislocation can be removed by virtually cutting the piece of material as indicated in Fig. 1.5. This leads to a purely plastic, incompatible deformation and well defined atom positions (Fig. 1.5, right). The width b of the opening gap represents the absolute value of the Burgers vector characterizing the dislocation. In general, the Burgers vector can be determined from the Burgers circuit (Frank, 1951), which is a closed atom-to-atom path in the deformed crystal (Fig. 1.5, left). In an undeformed, ordered crystal the same path is in general not closed, if it contains dislocations (Fig. 1.5, right). The closure-vector is the Burgers vector $\boldsymbol{b}$.

1.1.3 Regularization of Discrete Dislocations

The edge and screw dislocation stress fields (1.1) and (1.2) are singular at the dislocation center. This is a result of the localized plastic slip (displacement jump), which is bounded by the dislocation line. In some situations, it is useful to regularize the stresses by replacing the localized plastic slip by a smoothed version. This can for example be effectuated by first defining the plastic slip as delta distribution and then replacing the sharp delta by a smooth function, e.g., a Gaussian distribution with finite width (variance). As a consequence, the stresses are decreased to finite values close to the core. However, they can still be approximated by (1.1) and (1.2) at sufficient distances from the center, i.e., at radial distances which are much larger than the width of the Gaussian distribution.

Note carefully that the energy (1.3) will also decrease if the stress field is regularized, depending on the width of the regularized delta distribution. Hence, the smoothing procedure also changes the energy of the dislocation. This observation is also important in the context of certain continuum plasticity theories, which do not explicitly resolve individual dislocations. Instead, an averaged (or smoothed) description of the crystal kinematics is adopted, i.e., information is lost deliberately. It is important to note that this smoothing procedure requires special energetical considerations in order to accurately describe the energy of the system under consideration.

1.2 Dislocations in Applied Material Science

Since the 1930s, the discovery of dislocations initiated a significant improvement of the mechanical understanding of metal plasticity (e.g. Orowan, 1934; Polanyi, 1934; Taylor, 1934). One of the first important perceptions was that the macroscopically observed plastic deformation is not the result of the spontaneous sliding of whole atomic planes. This mechanism would require a significantly larger applied stress which is known as the theoretical strength being a multiple of the experimentally measured strength. Instead, the material resistance against plastic deformation is mainly determined by the stress required to trigger dislocation motion. However, besides dislocations additional origins of plastic deformations exist (e.g. formation of twins and grain boundary sliding).

1.2.1 Precipitation Hardening

Nowadays, the knowledge about dislocations is systematically exploited in the development and optimization of mechanical material properties. This is usually based on the introduction of various obstacles against dislocation motion. For example, the obstacles may be given by precipitates interacting with dislocations in various ways. If the crystallographic structure and orientation of the precipitates are coherent with the matrix, dislocations tend to cut the precipitates, if these are sufficiently small. However, dislocations normally bypass incoherent and large precipitates, e.g., by leaving their slip plane through climb or cross slip. Here, climb denotes dislocation motion normal to the slip plane due to diffusion of vacancies while cross slip can be effectuated by screw dislocations changing to another glide plane.
Another mechanism to bypass a precipitate is the Orowan mechanism, i.e., the dislocation bends around the impurity until it is fully surrounded and the dislocation can move on, leaving a so-called Orowan loop around the precipitate (Fig. 1.6).

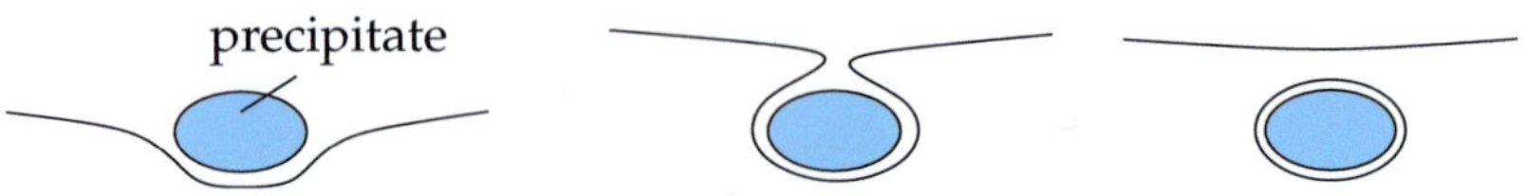

Figure 1.6: Illustration of the Orowan mechanism.

1.2.2 Substitutional and Interstitial Atoms – Dynamic Strain Aging

Besides precipitates, interstitial and substitutional atoms can be added to a material to increase its strength. The principal strengthening mechanism of substitutional atoms can be explained by the lattice distorsion around the atom which represents an energy barrier against dislocation motion.

In contrast, interstitial atoms additionally tend to move to the dislocation core by diffusion. Since the dislocation core is a center of dilatation, it represents an energetically favorable position for interstitials. When applied to an external shear stress, the pinned dislocation first needs to be torn away from the interstitials in order to generate plastic deformations. Macroscopically, this effect can often be observed by the existence of an upper and lower yield point in the stress strain diagram. When the upper yield point is reached, the deformation changes from elastic to plastic accompanied by a sudden stress drop representing the relieved dislocations. Simultateously, instead of being homogeneous, the plastic deformation usually starts at some small region of the specimen which subsequently expands until the whole specimen deforms plastically. Subsequently the material starts to harden.

After the deformation, the interstitials travel back to the dislocation cores leading once more to an increased strength. Depending on the temperature, this diffusion-controlled process may take some time. Under certain conditions, the diffusion speed competes with the dislocation velocity, leading to a dynamic pinning and unpinning of the dislocations during the deformation process. As a consequence, the microscopic strain rate sensitivity can become negative and localization phenomena become visible. This behavior is referred to as Portevain-Le Chatelier effect (PLC effect, e.g. Portevin and Le Chatelier, 1923) . It manifests itself in an oscillating stress strain diagram reflecting the competition of pinning and unpinning of dislocations (Fig. 1.7).

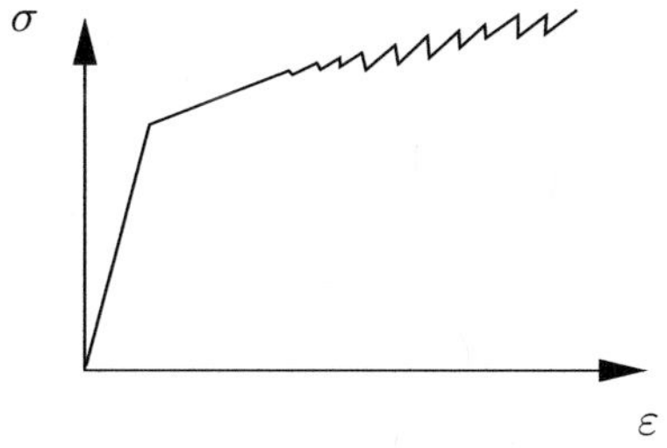

Figure 1.7: Illustration of the PLC effect.

1.2.3 Hall-Petch Effect

Grain boundaries represent another effective obstacle against dislocation motion. Therefore, the material strength can be increased by decreasing the grain size. This mechanism is referred to as the Hall-Petch effect (Hall, 1951; Petch, 1953). It is a favorable strengthening mechanism since it usually leads to an improved ductility of the material, which is usually desirable in actual applications. Dislocations interact with grain boundaries in manifold ways. For example, a grain boundary may serve as dislocation source or sink and dislocations may pile up in front of it.
If the grains are extremely fine, the main plastic deformation mechanism changes from dislocation motion to grain boundary sliding. This is accompanied by a reversal of the strengthening effect with decreasing grain size, known as the inverse Hall-Petch effect.

1.2.4 Single Crystals and Texture Strengthening

If a polycrystal is loaded in tension, several grains are always oriented such that the resolved shear stress takes large values in certain slip

systems. These grains usually initiate the plastic deformation.
Similarly, the tensile yield stress of a single crystal strongly depends on its orientation. This is exploited technically, e.g., in single crystalline turbine blades by optimizing the crystal orientation with respect to the load carrying capacity of the component under consideration.
Similarly, the strength of polycrystalline metals may be increased if the crystal orientation distribution is manipulated systematically, i.e., a texture is introduced for purpose.

1.2.5 Work Hardening

If a metallic specimen is deformed plastically, the number of dislocations and the total dislocation line lenght usually increase monotonously due to the presence of various sources of dislocation multiplication, like spiral sources or Frank-Read sources (Fig. 1.8) as well as dislocation interactions. The newly formed dislocations act as additional obstacles to other dislocations leading to an increased overall yield stress. This effect is called work hardening.

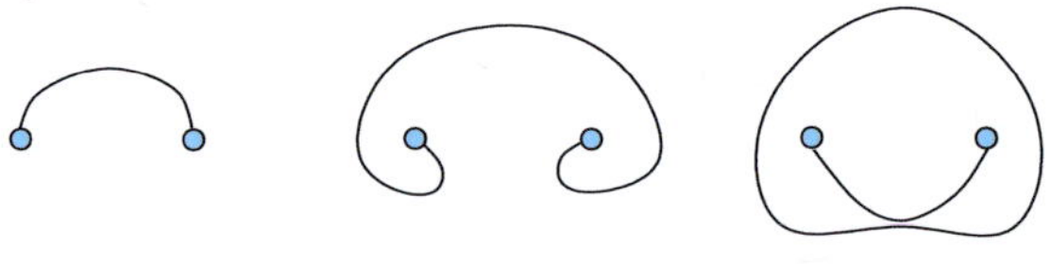

Figure 1.8: Frank-Read source.

The dislocation density of heavily deformed components can be reduced by recrystallisation. During this sort of heat treatment so-called recrystallization fronts move through the material. These interfaces separate regions with high and low dislocation densities. They move in the direction of the high-density region leaving an almost defect-free region behind, i.e., the atoms are reordered in the recrystallization front. The principal driving force of the interface motion is the energy difference between both aforementioned regions.

In heavily distorted metals, recrystalization allows for a significant refinement of the grains and a substantial Hall-Petch effect.

1.3 Continuum Description of Dislocations in Crystal Plasticity

Continuum crystal plasticity models allow to estimate the mechanical response of a body $\mathcal{B}$ subjected to a given set of boundary conditions. In general, the body may consist of several phases, at least some of which are considered to be crystals in the following. Usually, crystal plasticity models are applied to predict the bulk behavior of materials with microstructure (e.g. in a two-scale simulation) or to simulate micro components. In the small deformation context, one is interested in the deformation in terms of the displacement field $\boldsymbol{u}$ or the infinitesimal strain tensor $\boldsymbol{\varepsilon} = \mathrm{sym}(\boldsymbol{H})$. Here, $\boldsymbol{H} = \nabla \boldsymbol{u} = \partial_{x_j} u_i \boldsymbol{e}_i \otimes \boldsymbol{e}_j$ is the displacement gradient and $\{\boldsymbol{e}_1, \boldsymbol{e}_2, \boldsymbol{e}_3\}$ is an orthonormal basis.

1.3.1 Orowan's Equation

In continuum crystal plasticity, dislocations are usually described in terms of density measures and the individual dislocations are normally not resolved[1]. For example, one introduces the dislocation density ρ, which is interpreted as the "total line length per unit volume". Normally, the continuum crystal plasticity approach is useful in situations where the total dislocation line length in a considered volume element is much larger than the characteristic dimensions of the volume itself. In this case, it would be technically impossible to resolve each and every dislocation. Since real crystals usually have several slip systems, one introduces the dislocation density for each system separately. Here, the single slip situation is considered for simplicity.

[1] Discrete dislocations can also be described by continuum theories. However, here the term "continuum crystal plasticity" is used for theories which do not resolve individual dislocations.

Orowan (1934) motivated the following relation between the plastic slip rate and the dislocation density

$$\dot{\gamma} = \rho\, b\, \nu_{\text{avg}}, \tag{1.4}$$

where ν_{avg} is the line average of the dislocation velocity in the considered volume element. If a fraction of the dislocations is immobile, only the remaining, mobile part of the dislocation density is considered in Eq. (1.4).

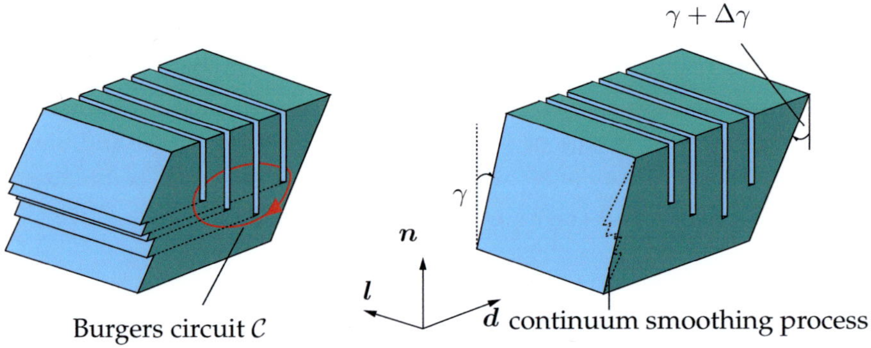

Figure 1.9: Discrete (localized) and continuum representation of the plastic slip γ.

The plastic slip γ can be interpreted as a smoothed version of the actual, localized plastic slip. This smoothing procedure is visualized schematically at the edges of the body in Fig. 1.9 (the smoothing of the dislocations is not visualized). Here, several edge dislocations have entered the material from the surface, leaving respective steps. These are not visible in the continuum crystal plasticity representation.

1.3.2 Plastic Distorsion and Dislocation Density Tensor

In addition to the aforementioned observations, Fig. 1.9 (right) illustrates that the presence of several dislocations of equal sign lead to a non-vanishing plastic slip gradient $\nabla\gamma$ in the continuum crystal plasticity representation. This relation will be discussed in more detail in the following. Therefore, the plastic distorsion is defined

$$\boldsymbol{H}^{\mathrm{p}} = \gamma \boldsymbol{d} \otimes \boldsymbol{n}. \tag{1.5}$$

In Fig. 1.9, the elastic lattice deformation has been relaxed by cutting the material as explained in Sect. 1.1.2. It should be mentioned that the material could have been cut in an infinite number of ways in order to elastically relax the lattice.

Fig. 1.9 shows that the plastic deformation is obviously incompatible. This can be seen from the Burgers circuit which, initially closed, is now interrupted at the cutting positions. In the given example, the value of the total closure or Burgers vector is given by $-4b$. It is important to note that this value coincides with the total amount of (discrete) plastic slip, which occurs at the Burgers circuit (at the points where the dislocation traces intersect with the Burgers circuit). This connection becomes particularly obvious if the dislocation traces are chosen as cutting positions. It also holds true, if additional edge dislocations with opposite sign or more complex dislocations with mixed character are surrounded by the Burgers circuit.

Motivated by these considerations, the continuum crystal plasticity analogon of the total or net Burgers vector $\boldsymbol{b}_{\mathrm{tot}}$ is defined as the following line integral

$$\boldsymbol{b}_{\mathrm{tot}} = \boldsymbol{d} \int_{\mathcal{C}} \gamma \boldsymbol{n} \cdot \mathrm{d}\boldsymbol{x} = \int_{\mathcal{C}} (\gamma \boldsymbol{d} \otimes \boldsymbol{n})\, \mathrm{d}\boldsymbol{x} = \int_{\mathcal{C}} \boldsymbol{H}^{\mathrm{p}}\, \mathrm{d}\boldsymbol{x}, \tag{1.6}$$

where $\mathcal{C}$ denotes an arbitrary Burgers circuit. Note that Eq. (1.6) also applies to the discrete case, if the localized plastic slip is inserted.

The application of Stokes' theorem leads to

$$\boldsymbol{b}_{\mathrm{tot}} = \int_{\mathcal{C}} \boldsymbol{H}^{\mathrm{p}}\, \mathrm{d}\boldsymbol{x} = \int_{A} \mathrm{curl}^{\mathrm{T}}(\boldsymbol{H}^{\mathrm{p}})\, \mathrm{d}\boldsymbol{a}, \tag{1.7}$$

with $\mathrm{curl}(\boldsymbol{H}^{\mathrm{p}}) = \epsilon_{ijk} \partial_{x_i} H^{\mathrm{p}}_{lj} \boldsymbol{e}_k \otimes \boldsymbol{e}_l$, where ϵ_{ijk} is the permutation symbol. The area A is an arbitrary area bounded by $\mathcal{C} = \partial A$. The direction of $\mathrm{d}\boldsymbol{a}$

is given by the normal of A at the corresponding position. Eq. (1.7) motivates the interpretation of the tensor

$$\boldsymbol{\alpha} = \mathrm{curl}^{\mathrm{T}}(\boldsymbol{H}^{\mathrm{p}}) \tag{1.8}$$

as a dislocation density tensor (Kondo, 1952; Nye, 1953; Bilby et al., 1955; Kröner, 1958). It may be interpreted as a generalized "Burgers vector density per unit area".
If exclusively edge dislocations are present (like in Fig. 1.9), the dislocation density tensor reads

$$\boldsymbol{\alpha} = -(\boldsymbol{d} \cdot \nabla\gamma)\boldsymbol{d} \otimes \boldsymbol{l} = \rho_{\vdash}\boldsymbol{d} \otimes \boldsymbol{l}, \tag{1.9}$$

where $\rho_{\vdash} = -(\boldsymbol{d} \cdot \nabla\gamma)$ may be interpreted as scalar edge dislocation density.
It should be noted that the identity $\rho = \rho_{\vdash}/b$ holds only if the dislocations share the same sign and direction. This is not the case, for example, if the amount of positive dislocations equals that of negative dislocations. Then, $\rho_{\vdash}$ is zero and ρ takes a finite value.
In general, the total dislocation density is decomposed into a geometrically necessary part and a remaining part denoted by "statistically stored dislocations" (SSDs). The portion of the underlying dislocations which is related to $\boldsymbol{\alpha}$ represents the geometrically necessary dislocations (GNDs, Ashby (1970)).
The fact that the curl of $\boldsymbol{H}^{\mathrm{p}}$ does in general not vanish is equivalent to the statement that $\boldsymbol{H}^{\mathrm{p}}$ itself cannot be derived from a potential. This means that, generally, there is no plastic displacement field, i.e., the plastic deformation is usually incompatible as discussed in Sect. 1.1.2. Therefore, the dislocation density tensor is often referred to as a measure of incompatibility.
SSDs are dislocations which trap each other randomly and do therefore not contribute to plastic strain gradients (Bortoloni and Cermelli, 2000).

1.4 A Classical Continuum Crystal Plasticity Framework for Single Slip

1.4.1 Elastic Properties

It is assumed that the total displacement gradient decomposes additively into an elastic part $\boldsymbol{H}^{\mathrm{e}}$ and a plastic part $\boldsymbol{H}^{\mathrm{p}}$,

$$\boldsymbol{H} = \boldsymbol{H}^{\mathrm{e}} + \boldsymbol{H}^{\mathrm{p}}. \tag{1.10}$$

The elastic and plastic strains can be defined as $\boldsymbol{\varepsilon}^{\mathrm{e}} = \mathrm{sym}(\boldsymbol{H}^{\mathrm{e}})$ and $\boldsymbol{\varepsilon}^{\mathrm{p}} = \mathrm{sym}(\boldsymbol{H}^{\mathrm{p}})$, respectively. Hence, the additive decomposition also applies to the infinitesimal strain tensor

$$\boldsymbol{\varepsilon} = \mathrm{sym}(\boldsymbol{H}) = \boldsymbol{\varepsilon}^{\mathrm{e}} + \boldsymbol{\varepsilon}^{\mathrm{p}}. \tag{1.11}$$

A typical constitutive assumption for the Cauchy stress is Hooke's law

$$\boldsymbol{\sigma} = \mathbb{C}[\boldsymbol{\varepsilon}^{\mathrm{e}}], \tag{1.12}$$

with the elastic stiffness tensor $\mathbb{C}$, which is assumed to have the usual properties (major and minor symmetry as well as positive definiteness). Defining the volumetric elastic energy density by

$$W_{\mathrm{e}} = \frac{1}{2}\boldsymbol{\varepsilon}^{\mathrm{e}} \cdot \mathbb{C}[\boldsymbol{\varepsilon}^{\mathrm{e}}], \tag{1.13}$$

the stress power is given by

$$\boldsymbol{\sigma} \cdot \dot{\boldsymbol{\varepsilon}} = \boldsymbol{\sigma} \cdot \dot{\boldsymbol{\varepsilon}}^{\mathrm{e}} + \boldsymbol{\sigma} \cdot \dot{\boldsymbol{\varepsilon}}^{\mathrm{p}} = \partial_{\boldsymbol{\varepsilon}^{\mathrm{e}}} W_{\mathrm{e}} \cdot \dot{\boldsymbol{\varepsilon}}^{\mathrm{e}} + \dot{\gamma}\boldsymbol{\sigma} \cdot (\boldsymbol{d} \otimes \boldsymbol{n}) = \dot{W}_{\mathrm{e}} + \tau\,\dot{\gamma}. \tag{1.14}$$

Here, $\tau = \boldsymbol{\sigma} \cdot (\boldsymbol{d} \otimes \boldsymbol{n})$ denotes the resolved shear stress of the considered slip system.

1.4.2 Viscoplastic Formulation

In general, the plastic response of real crystals is rate-dependent, i.e., the stress response depends on the applied strain rate. Motivated by the typical power-law type creep behavior of metals, these are often of the form (Hutchinson, 1976)

$$\dot{\gamma} = \operatorname{sgn}(\tau)\, \dot{\gamma}_0 \left| \frac{\tau}{\tau^{\mathrm{C}}} \right|^p . \tag{1.15}$$

Here, $\dot{\gamma}_0$ is a reference shear rate and p is a strain rate sensitivity parameter. For large values of p, Eq. (1.15) approximates a rate-independent material with the critical resolved shear stress τ^{C}. Alternatively, the following overstress approach is often used

$$\dot{\gamma} = \operatorname{sgn}(\tau)\, \dot{\gamma}_0 \left\langle \frac{|\tau| - \tau^{\mathrm{C}}}{\tau^{\mathrm{D}}} \right\rangle^p , \tag{1.16}$$

where τ^{D} is called drag stress and $\langle \bullet \rangle = (\bullet + | \bullet |)/2 = \max\{\bullet, 0\}$.
Creep is highly controlled by dislocations bypassing obstacles by climb, i.e., the creep behavior strongly depends on the diffusion properties of vacancies. Therefore, the strain rate sensitivity is a function of the temperature.
A rate-independent model can be considered a reasonable approximation in many applications. However, it is well known that in some situations, the plastic slip rates may not be unique in the rate-independent multislip case (Taylor, 1938; Kocks, 1970). This requires a special numerical treatment (see, e.g., Miehe and Schröder, 2001). Therefore, rate-dependent models (e.g. Perzyna, 1971) are often applied.

1.4.3 Work Hardening Models

Equation (1.14) illustrates that the resolved shear stress τ is power conjugate to $\dot{\gamma}$, i.e., the product $\tau\dot{\gamma}$ contributes to the mechanical power (1.14). Hence, τ can be interpreted as plastic driving force. Indeed, Schmid and Boas (1935) found experimentally that many metals yield when the resolved shear stress τ reaches a critical value τ^{C}. This relation is referred to as 'Schmid law'. Usually, this is a good approximation for face centered cubic (FCC) metals, while body centered cubic (BCC) metals show more complex slip modes and partially violate the Schmid law (see, e.g., Havner, 1992; Yalcinkaya et al., 2008).

The critical resolved shear stress τ^{C} depends on several parameters, for example the temperature or the concentration of interstitial atoms. In general, the critical resolved shear stress evolves during plastic deformation processes due to work-hardening. Phenomenological hardening models have been proposed, for example, by Taylor (1938), Koiter (1953) or Hill (1966) who proposed linear hardening relations for the multislip case. These relate the critical resolved shear stresses τ_α^{C} to the slip rates of the individual slip systems via the evolution law

$$\dot{\tau}_\alpha^{\mathrm{C}} = \sum_\beta h_{\alpha\beta} |\dot{\gamma}_\beta|, \tag{1.17}$$

where $h_{\alpha\beta}$ are the hardening moduli.
Kocks (1970) further investigated the relation between self- and latent hardening and found that the latent hardening is usually stronger than the self-hardening rate (e.g. by a factor up to 1.4). This can be formalized by

$$h_{\alpha\beta} = H[q_{\mathrm{l}} + (1 - q_{\mathrm{l}})\delta_{\alpha\beta}], \tag{1.18}$$

with $q_{\mathrm{l}} \in [1, 1.4]$ (see also Cuitiño and Ortiz, 1992). In order to account for the hardening nonlinearity, the modulus H is often taken as a function of an equivalent plastic strain γ_{eq} which may, for example, be defined by

$$\gamma_{\mathrm{eq}} = \sum_\alpha \int_0^t |\dot{\gamma}_\alpha| \, \mathrm{d}\tilde{t}. \tag{1.19}$$

Alternative (but closely related) hardening models are based on heuristic dislocation density evolution equations of the saturation type (Gillis and Gilman, 1965; Essmann and Rapp, 1973). For example, for the single slip case, these relations often take the following form

$$\dot{\rho} = c_\rho (1 - \rho/\rho_{\mathrm{sat}}) |\dot{\gamma}|, \tag{1.20}$$

where c_ρ and ρ_{sat} are model parameters. Typically, the critical shear stress is estimated by the relation (Taylor, 1934)

$$\tau^{\mathrm{C}} = \tau_0^{\mathrm{C}} + a_{\mathrm{T}} G b \sqrt{\rho}, \tag{1.21}$$

where a_{T} is a constant of order one. For multislip, Eq. (1.21) must be generalized to account for multiple dislocation densities (see, e.g., Franciosi and Zaoui, 1982).
Recent works on gradient plasticity apply equations of the form (1.20) as evolution equations of the SSD-density (Evers et al., 2004b; Becker, 2006).

1.5 Gradient Crystal Plasticity Based on Scalar GND-Densities

1.5.1 Motivation

One objective of the work at hand is the investigation and formulation of phenomenological gradient plasticity theories. In this context, geometrically necessary dislocations (GNDs) play a major role. Therefore, this section extends the geometrically linear kinematical description of geometrically necessary dislocations outlined in Sect. 1.3.2. In addition, a prominent phenomenological gradient crystal plasticity model based on an energetic approach is shortly reviewed. A simple two-dimensional shear example will be discussed in order to illustrate the typical mechanical behavior of this widespread type of gradient plasticity model.

1.5.2 Different Representations of Geometrically Necessary Dislocations

The classical kinematical framework of geometrically linear elasto-plasticity has been introduced in Sect. 1.4. It is based on the strain $\varepsilon = \mathrm{sym}(\boldsymbol{H})$ as the symmetric part of the displacement gradient $\boldsymbol{H}$ and the additive decompositions $\boldsymbol{H} = \boldsymbol{H}^{\mathrm{e}} + \boldsymbol{H}^{\mathrm{p}}$ as well as $\varepsilon = \varepsilon^{\mathrm{e}} + \varepsilon^{\mathrm{p}}$ of these tensors into elastic and plastic parts.
In gradient plasticity, this kinematical framework is extended. The principal motivation of this extension are size effects exhibited by metals, when the length scale of the system under consideration is in the order

of microns. Size effects have been observed in many experiments using micro-specimens (see the introduction for an overview). As a general rule it was found that the smaller the system size, the stronger the mechanical response. Additional examples are given by the Hall-Petch effect and precipitation hardening, where it has been found that the strengthening does not only depend on the precipitates' volume fraction but also on their size.

Classical continuum plasticity models without internal length scale are unable to reproduce the observed size-dependence. Some experiments (e.g. Fleck et al., 1994) suggest a correlation between the size effect and the plastic spatial variation, i.e., deformation inhomogeneity. As a consequence, phenomenological models have been proposed which assume the material strength to depend on the gradient of some plastic strain or strain-like variable and thereby introduce an internal length scale into the theory. Many of these gradient extended models are based on Nye's dislocation density tensor (Eq. (1.8))

$$\boldsymbol{\alpha} = \mathrm{curl}^{\mathrm{T}} \boldsymbol{H}^{\mathrm{p}} = -\mathrm{curl}^{\mathrm{T}} \boldsymbol{H}^{\mathrm{e}}. \tag{1.22}$$

Here, it has been exploited that the curl of the displacement gradient $\boldsymbol{H} = \boldsymbol{H}^{\mathrm{e}} + \boldsymbol{H}^{\mathrm{p}}$ vanishes due to the compatibility of the displacement gradient $\boldsymbol{H}$.

Relation (1.22) has been discussed for the case of single slip in Sect. 1.3.2. By superposition, it can also be applied to the geometrically linear multislip case based on the following representation of the plastic distorsion

$$\boldsymbol{H}^{\mathrm{p}} = \sum_{\alpha=1}^{N} \gamma_\alpha \boldsymbol{d}_\alpha \otimes \boldsymbol{n}_\alpha, \tag{1.23}$$

where N is the number of slip systems, γ_α denotes the plastic slip on slip system α and the vectors $\boldsymbol{d}_\alpha$ and $\boldsymbol{n}_\alpha$ are the slip directions and slip plane normals, respectively. Eq. (1.23) is a superposition of the single slip case (1.5), which is only valid in the geometrically linear regime.

Inserting the representation (1.23) of $\boldsymbol{H}^{\mathrm{p}}$ into Eq. (1.22) yields

$$\boldsymbol{\alpha} = \sum_{\alpha=1}^{N} \boldsymbol{d}_\alpha \otimes (\rho_\vdash^\alpha \boldsymbol{l}_\alpha + \rho_\odot^\alpha \boldsymbol{d}_\alpha). \tag{1.24}$$

Here, the scalar edge and screw dislocation densities $\rho_\vdash^\alpha = -\boldsymbol{d}_\alpha \cdot \nabla\gamma_\alpha$, $\rho_\odot^\alpha = \boldsymbol{l}_\alpha \cdot \nabla\gamma_\alpha$ as well as the lattice vectors $\boldsymbol{l}_\alpha = \boldsymbol{n}_\alpha \times \boldsymbol{d}_\alpha$ have been introduced.

1.5.3 A Gradient Plasticity Hardening Model

Gradient plasticity theories have in common that they model the material behavior to be dependent on geometrically necessary dislocations or similar quantities. However, this can be done and was actually done in various ways (see the introduction for an overview). The probably most popular approach introduces the GNDs as additional arguments of the energy density (e.g. Steinmann, 1996; Cermelli and Gurtin, 2002; Berdichevsky, 2006a).

In the following, a typical computational example is discussed (see, e.g., Wulfinghoff and Böhlke, 2010). This example illustrates typical dislocation microstructures which arise from the aforementioned widespread energy based approach. In addition, the model's features are compared to a theory without internal length scale. Note that the theory is not presented in detail here (for additional information see Gurtin et al., 2007), since the purpose of this section is a first illustration of some key features, including the notion of field equations, generalized stresses and the overall behavior of gradient plasticity models instead of a detailed theoretical examination which is carried out in the subsequent chapters.

The field equations may be derived from the principle of virtual power. Here, they are assumed to be given by the linear momentum balance $\mathrm{div}(\boldsymbol{\sigma}) = \boldsymbol{0}$ and additional microforce balance equations for the dissipative force $\tau_\alpha^{\mathrm{d}} = \tau_\alpha - \tau_\alpha^{\mathrm{B}}$ in the αth slip system, where $\boldsymbol{\sigma}$ denotes the Cauchy stress tensor. The dissipative forces τ_α^{d} determine the (volumetric) dissipation density $\mathcal{D} = \sum_{\alpha} \tau_\alpha^{\mathrm{d}} \dot{\gamma}_\alpha$. Moreover, $\tau_\alpha = \boldsymbol{\sigma} \cdot (\boldsymbol{d}_\alpha \otimes \boldsymbol{n}_\alpha)$ is the resolved shear stress of slip system α and

$$\tau_\alpha^{\mathrm{B}} = -\operatorname{div}(\boldsymbol{\xi}_\alpha) \tag{1.25}$$

can be interpreted as a backstress resulting from the presence of GNDs. The components of the generalized stresses $\boldsymbol{\xi}_\alpha$ are here assumed energetically conjugate to the scalar edge and screw dislocation densities $\rho_\vdash^\alpha$ and $\rho_\odot^\alpha$ (Eq. (1.24)) as explained in the following.

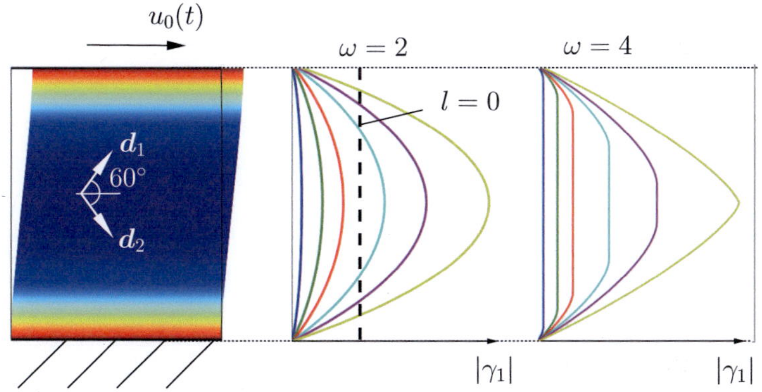

Figure 1.10: Plastic shear strain evolution for $\omega = 2$ and $\omega = 4$

The elastic energy is supplemented by an additional phenomenological term W_{g}, i.e., $W = W_{\mathrm{e}}(\boldsymbol{\varepsilon}, \boldsymbol{\varepsilon}^{\mathrm{p}}) + W_{\mathrm{g}}(\rho_\vdash^\alpha, \rho_\odot^\alpha)$, also denoted as defect energy (Gurtin et al., 2007). For simplicity, no isotropic hardening is considered. The generalized stresses

$$\boldsymbol{\xi}_\alpha = \frac{\partial W_{\mathrm{g}}}{\partial \nabla \gamma_\alpha} = -\frac{\partial W_{\mathrm{g}}}{\partial \rho_\vdash^\alpha} \boldsymbol{d}_\alpha + \frac{\partial W_{\mathrm{g}}}{\partial \rho_\odot^\alpha} \boldsymbol{l}_\alpha \tag{1.26}$$

are also denoted as microstresses. The aforementioned expressions

$$\tau_\alpha + \operatorname{div}(\boldsymbol{\xi}_\alpha) - \tau_\alpha^{\mathrm{d}} = 0 \tag{1.27}$$

represent microforce balance equations, which supplement the linear momentum balance.

To generate a closed set of equations, thermodynamically consistent constitutive functions are assumed for the dissipative stresses $\tau_\alpha^{\mathrm{d}}(\dot{\gamma}_\alpha)$, the strain energy $W_{\mathrm{e}}(\boldsymbol{\varepsilon}, \boldsymbol{\varepsilon}^{\mathrm{p}})$ and the defect energy $W_{\mathrm{g}}(\rho_\vdash^\alpha, \rho_\odot^\alpha)$. A quadratic approach is chosen for $W_{\mathrm{e}}(\boldsymbol{\varepsilon}, \boldsymbol{\varepsilon}^{\mathrm{p}})$, the dissipative stresses $\tau_\alpha^{\mathrm{d}}(\dot{\gamma}_\alpha)$ are assumed to be given by a power law. Two decoupled approaches

$$W_{\mathrm{g}} = c_\omega \sum_\alpha |\rho_\vdash^\alpha l|^\omega \tag{1.28}$$

with $\omega = 2$ (Gurtin et al., 2007) and $\omega = 4$ are compared (c_ω is another material property and l is an internal length scale parameter).

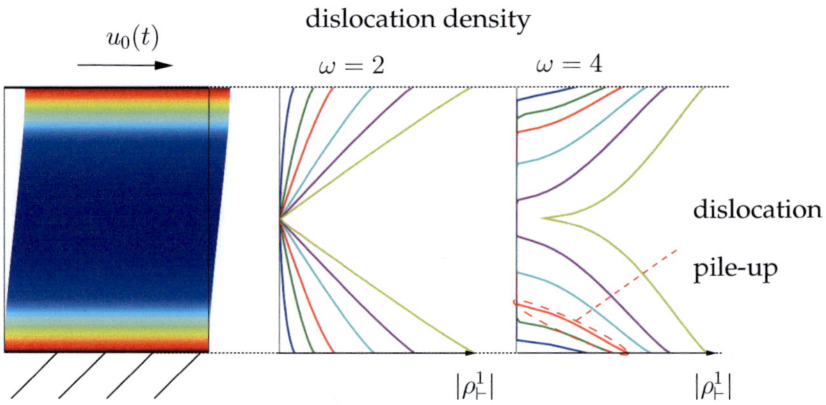

Figure 1.11: Dislocation pile-up evolution for $\omega = 2$ and $\omega = 4$

The analyzed periodic plane strain problem involves two slip systems, the slip directions of which are $\pm 60°$ with respect to the horizontal. The displacement at the top is prescribed. The top and bottom are assumed impenetrable for dislocations. This can be accounted for by setting the Dirichlet boundary conditions involving the plastic slips to zero. Physically, these so-called micro-hard boundary conditions can be interpreted as passivated surface layers or idealized grain boundaries.
The results (see Figs. 1.10 and 1.11) show, that the first ansatz ($\omega = 2$) leads to smooth strain curves (Fig. 1.10, center) and dislocation pile-ups (Fig. 1.11, center). These pile-ups generate a long-range influence

from the boundary on the bulk behavior. For comparison, the result of the classical theory without internal length scale, i.e. $l = 0$ is also depicted in Fig. 1.10. In this case, no dislocation pile ups emerge and the deformation is homogeneous. The ansatz $\omega = 4$ generates a clearer distinction between bulk and boundaries than $\omega = 2$.
The size effect due to the quadratic defect energy extension is depicted qualitatively in Fig. 1.12. The overall hardening is size-dependent, the yield strength is not. In addition, the diagram shows the model response for vanishing defect energy (i.e. $l = 0$) which is comparable to the model response for an infinite layer thickness.

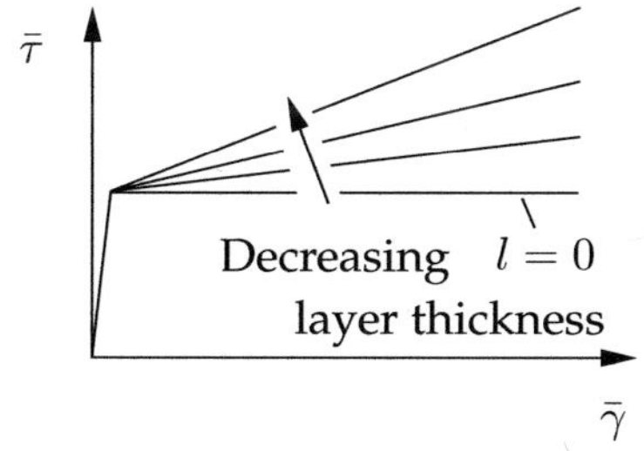

Figure 1.12: Average shear stress-strain-diagram for the quadratic defect energy model. The smaller the layer thickness the stronger the overall hardening.

Chapter 2

Application of Gradient Plasticity to a Laminate Material – Inspection of Energies

The chapter at hand is largely taken from Wulfinghoff et al. (2014). It introduces the framework of gradient crystal plasticity based on the dislocation density tensor. Since the framework is exploited in the subsequent sections, the derivations are given in detail.

2.1 Gradient Crystal Plasticity Based on the Dislocation Density Tensor

The extension of crystal plasticity based on GNDs is not unique. For example, the constitutive equations may account for scalar edge and screw densities (as illustrated in Sect. 1.5). These are defined for each slip system and interact through constitutive models (e.g. Gurtin et al., 2007). Alternatively, the extension may be based on the full dislocation density tensor (Eq. (1.22)). In this case, the interaction is implicitly defined (although little is known about the accuracy of this approach). Besides the Cauchy stress $\boldsymbol{\sigma}$, additional generalized stresses $\boldsymbol{s}$ and $\boldsymbol{M}$ are introduced, which are in general nonsymmetric. These are second-order tensors being work-conjugate to $\dot{\boldsymbol{H}}^{\mathrm{p}}$ and $\dot{\boldsymbol{\alpha}}$, respectively. The virtual power of this internal force system is assumed to be given by

$$\delta\mathcal{P}_{\mathrm{int}} = \int_{\mathcal{B}} \delta p_{\mathrm{int}}\,\mathrm{d}v = \int_{\mathcal{B}} \left(\boldsymbol{\sigma}\cdot\delta\dot{\boldsymbol{\varepsilon}} + \boldsymbol{s}\cdot\delta\dot{\boldsymbol{H}}^{\mathrm{p}} + \boldsymbol{M}\cdot\delta\dot{\boldsymbol{\alpha}}\right)\mathrm{d}v, \qquad (2.1)$$

with $\delta\dot{\boldsymbol{\varepsilon}} = \mathrm{sym}(\nabla\delta\dot{\boldsymbol{u}})$ and $\delta\dot{\boldsymbol{\alpha}} = \mathrm{curl}^{\mathrm{T}}(\delta\dot{\boldsymbol{H}}^{\mathrm{p}})$.
In absence of volume forces, the virtual power of the external forces is

given by

$$\delta \mathcal{P}_{\text{ext}} = \int_{\partial \mathcal{B}_t} \bar{\boldsymbol{t}} \cdot \delta \dot{\boldsymbol{u}} \, \text{da} + \int_{\partial \mathcal{B}_m} \bar{\boldsymbol{m}} \cdot \delta \dot{\boldsymbol{H}}^{\text{p}} \, \text{da}, \tag{2.2}$$

where the integration domains represent Neumann-type boundaries with prescribed tractions $\bar{\boldsymbol{t}}$ and microtractions $\bar{\boldsymbol{m}}$. The evaluation of the principle of virtual power under static conditions,

$$\delta \mathcal{P}_{\text{int}} = \delta \mathcal{P}_{\text{ext}}, \tag{2.3}$$

gives, in combination with the chain rule and Gauss' theorem, the field equations at regular points

$$\text{div}\,(\boldsymbol{\sigma}) = \boldsymbol{0}, \quad \boldsymbol{s} + \text{curl}^{\text{T}}(\boldsymbol{M}) = \boldsymbol{0}. \tag{2.4}$$

In addition, the principle yields the following Neumann-type boundary conditions on $\partial \mathcal{B}_t$ and $\partial \mathcal{B}_m$

$$\bar{\boldsymbol{t}} = \boldsymbol{\sigma} \boldsymbol{n}, \quad \bar{\boldsymbol{m}} = \boldsymbol{M} \hat{\boldsymbol{n}}, \tag{2.5}$$

with

$$\hat{\boldsymbol{n}} = -\boldsymbol{\epsilon} \boldsymbol{n} = -(\epsilon_{ijk} \boldsymbol{e}_i \otimes \boldsymbol{e}_j \otimes \boldsymbol{e}_k)(n_l \boldsymbol{e}_l) = -\epsilon_{ijk} n_k \boldsymbol{e}_i \otimes \boldsymbol{e}_j. \tag{2.6}$$

The power density of the internal forces reads

$$p_{\text{int}} = \boldsymbol{\sigma} \cdot \dot{\boldsymbol{\varepsilon}} + \boldsymbol{s} \cdot \dot{\boldsymbol{H}}^{\text{p}} + \boldsymbol{M} \cdot \dot{\boldsymbol{\alpha}}. \tag{2.7}$$

Up to this point, the theory is quite general, since it may be applied to single as well as polycrystal plasticity. In the following, the work will focus on single crystal plasticity, where the plastic distorsion is given by the additional kinematical relation (1.23)

$$\boldsymbol{H}^{\text{p}} = \sum_{\alpha} \gamma_\alpha \boldsymbol{d}_\alpha \otimes \boldsymbol{n}_\alpha, \tag{2.8}$$

where $\boldsymbol{d}_\alpha$ and $\boldsymbol{n}_\alpha$ denote the slip directions and slip plane normals, respectively. The index α runs over all slip systems.
In addition, an equivalent plastic strain γ_{eq} is introduced. One possible definition is given by Eq. (1.19)

$$\gamma_{\text{eq}} = \int_0^t \sum_{\alpha} |\dot{\gamma}_\alpha| \, \text{d}\tilde{t}. \tag{2.9}$$

It is assumed that the volumetric stored energy density has the form

$$W = W_{\mathrm{e}} + W_{\mathrm{g}} + W_{\mathrm{h}}, \tag{2.10}$$

with $W_{\mathrm{e}} = (\boldsymbol{\varepsilon} - \boldsymbol{\varepsilon}^{\mathrm{p}}) \cdot \mathbb{C}[\boldsymbol{\varepsilon} - \boldsymbol{\varepsilon}^{\mathrm{p}}]/2$. The expressions W_{h} and W_{g} are assumed to be functions of $\hat{\mu}$ and $\boldsymbol{\alpha}$, respectively. Here, $\hat{\mu} = (\mu_1, \mu_2, ..., \mu_{N_\mu})$ denotes a vector of internal history variables. The functions are assumed normalized, i.e., $W_{\mathrm{h}}(\hat{0}) = 0$ and $W_{\mathrm{g}}(\mathbf{0}) = 0$. Isotropic hardening is accounted for by W_{h}, while W_{g} models size effects. The evolution of $\hat{\mu}$ is assumed to be given by the rate-independent approach

$$\dot{\mu}_\alpha = \sum_\beta f_{\alpha\beta}(\hat{\mu})\,|\dot{\gamma}_\beta| \quad \text{with} \quad f_{\alpha\beta}(\hat{\mu}) \geq 0. \tag{2.11}$$

The functions $f_{\alpha\beta}$ determine which specific hardening model is applied. For example, if $\hat{\mu} = (\gamma_{\mathrm{eq}})$, it follows that $f_{1\beta} = 1$.
In this chapter, the following two approaches will be investigated

$$W_{\mathrm{g}}^{1} = cGb\|\boldsymbol{\alpha}\|, \quad W_{\mathrm{g}}^{\ln} = c_0\|\boldsymbol{\alpha}\| \ln \frac{\|\boldsymbol{\alpha}\|}{\alpha_0}, \tag{2.12}$$

where c is a constant of order unity, G is the macroscopic shear modulus, b is the Burgers vector, α_0 is a constant and c_0 is given by

$$c_0 = \frac{Gbc_1}{2\pi(1-v)}, \tag{2.13}$$

where v is Poisson's ratio and c_1 is of order unity (in this work $c_1 = 1$). The Euclidean norm of the dislocation density tensor is defined as: $\|\boldsymbol{\alpha}\| = \sqrt{\boldsymbol{\alpha} \cdot \boldsymbol{\alpha}}$.
The rank-one energy W_{g}^{1} can be motivated by simple line tension arguments as follows. In many situations, the total elastic energy of a real crystal is well represented by W_{e}. However, the stored elastic energy around dislocations seems to be partially missing in W_{e}, since the regions close to the dislocation cores are not explicitly resolved in the continuum model (see the discussion in Sect. 1.1.3). In addition, one might assume that the respective energies of a set of statistically stored dislocations (SSDs) may be negligible, since SSDs tend to form

di- and multipoles. This implies a mutual screening of the individual stress fields and lower energies. Finally, it may be assumed that all other dislocation interactions are accounted for in W_e. Then, the remaining energy that needs to be accounted for is the stored elastic energy around geometrically necessary dislocations. The associated energy density is proportional to the amount of GNDs, since the interactions are assumed to be already represented by W_e. This explains the form of W_g^1 in Eq. (2.12). For additional arguments see Ortiz and Repetto (1999) and Hurtado and Ortiz (2012, 2013).

The logarithmic energy $W_g^{\ln}$ (Eq. (2.12)) is motivated by the form of the associated backstress (Forest and Guéninchault, 2013). It turns out, that the approach $W_g^{\ln}$ leads to a backstress which is formally close to the one derived in the statistical theory of Groma et al. (2003)[2], given by

$$-\frac{Gc_1}{2\pi(1-\upsilon)\rho}\partial^2_{x_1}\gamma \tag{2.14}$$

for a single slip situation with slip direction $\boldsymbol{e}_1$. Here, ρ denotes the total dislocation density. In the two-dimensional single slip regime, the backstress involves the Laplacian of the plastic slip, as postulated by Aifantis (1987). However, the internal length scale is not interpreted as a material constant but determined by the dislocation microstructure, if $W_g^{\ln}$ is applied. This point is discussed in detail in Sect. 2.2.2.

The subsequent sections investigate the features of the rank-one and logarithmic energies W_g^1 and $W_g^{\ln}$, respectively. Since the logarithmic energy is neither smooth nor convex, a regularization will also be discussed.

The stresses $\boldsymbol{\sigma}$ and $\boldsymbol{M}$ are assumed to be energetic, i.e.

$$\boldsymbol{\sigma} = \partial_{\varepsilon} W, \quad \boldsymbol{M} = \partial_{\alpha} W. \tag{2.15}$$

The dissipation inequality can be shown to be given by

$$\mathcal{D} = p_{\text{int}} - \dot{W} = \sum_{\alpha}(\tau_{\alpha}^{\text{eff}} - q_{\alpha}\,\text{sgn}\,(\dot{\gamma}_{\alpha}))\,\dot{\gamma}_{\alpha} \geq 0 \tag{2.16}$$

[2] Since Groma's work represents a two-dimensional single slip theory, this comparison is made for that situation.

with $\tau_\alpha^{\text{eff}} = (\boldsymbol{\sigma} + \boldsymbol{s}) \cdot (\boldsymbol{d}_\alpha \otimes \boldsymbol{n}_\alpha)$ and $q_\alpha = \Sigma_\beta f_{\alpha\beta} \partial_{\mu_\beta} W_{\text{h}}$. A possible flow rule, satisfying the dissipation inequality (2.16), is given by the following power law

$$\dot{\gamma}_\alpha = \operatorname{sgn}\left(\tau_\alpha^{\text{eff}}\right) \dot{\gamma}_0 \left\langle \frac{|\tau_\alpha^{\text{eff}}| - (\tau_0^{\text{C}} + q_\alpha)}{\tau^{\text{D}}} \right\rangle^p . \tag{2.17}$$

Here, τ_0^{C} is the initial yield stress, $\dot{\gamma}_0$ is a reference shear rate, p is the strain rate sensitivity and τ^{D} is a drag stress.
If the stored energy is not differentiable at $\boldsymbol{\alpha} = \boldsymbol{0}$, the symbol ∂ in Eq. $(2.15)_2$ is interpreted as a sub-differential operator (see, e.g., Han and Reddy, 2013), i.e.

$$\boldsymbol{M}\Big|_{\boldsymbol{\alpha}=\boldsymbol{0}} \in \{\boldsymbol{M} : W_{\text{g}}(\boldsymbol{\alpha}) - \boldsymbol{M} \cdot \boldsymbol{\alpha} \geq 0 \, \forall \, \boldsymbol{\alpha}\}. \tag{2.18}$$

This can be interpreted as follows. If the stress $\boldsymbol{M}$ is applied at a material point, $\boldsymbol{\alpha}$ will take a value which minimizes the expression $W_{\text{g}}(\boldsymbol{\alpha}) - \boldsymbol{M} \cdot \boldsymbol{\alpha}$. For small values of $\boldsymbol{M}$, the minimum is given by $\boldsymbol{\alpha} = \boldsymbol{0}$. However, for sufficiently large values of $\boldsymbol{M}$, the value of $\boldsymbol{\alpha}$ can be determined from the stationarity condition $\boldsymbol{M} = \partial_\alpha W_{\text{g}}$.

Example: If W_{g} is given by $W_{\text{g}}^1 = cGb\|\boldsymbol{\alpha}\|$, it follows that $\boldsymbol{\alpha} = \boldsymbol{0}$ if

$$\boldsymbol{M} \cdot \tilde{\boldsymbol{\alpha}} \leq W_{\text{g}}^1(\tilde{\boldsymbol{\alpha}}) = cGb\|\tilde{\boldsymbol{\alpha}}\| \quad \forall \tilde{\boldsymbol{\alpha}} \tag{2.19}$$

$$\Leftrightarrow \boldsymbol{M} \cdot \tilde{\boldsymbol{\alpha}} \leq \|\boldsymbol{M}\|\|\tilde{\boldsymbol{\alpha}}\| \leq cGb\|\tilde{\boldsymbol{\alpha}}\| \quad \forall \tilde{\boldsymbol{\alpha}}. \tag{2.20}$$

Hence, it is found that

$$\begin{cases} \boldsymbol{M} \in \{\boldsymbol{M} : \varphi_{\text{g}}^1(\boldsymbol{M}) \leq 0\}, \text{ if } \boldsymbol{\alpha} = \boldsymbol{0} \\ \boldsymbol{M} = cGb\dfrac{\boldsymbol{\alpha}}{\|\boldsymbol{\alpha}\|}, \text{ else.} \end{cases} \tag{2.21}$$

with $\varphi_{\text{g}}^1(\boldsymbol{M}) = \|\boldsymbol{M}\| - cGb$.
Remark: Note that the generalized stress $\boldsymbol{M}$ can be computed uniquely from $\boldsymbol{\alpha}$ only if $\boldsymbol{\alpha} \neq \boldsymbol{0}$. This makes analytical solutions as well as the numerical implementation difficult. The same problem arises in rate-independent rigid plasticity, where the stress can only be computed from the plastic slip rate if this is non-vanishing. Possible regularization techniques include the introduction of small elastic strains or the

approximation of the rate-independent model by, e.g., a power law with a large rate-sensitivity. The introduction of elastic strains usually implies better convergence properties. Therefore, the work at hand introduces an analogue regularization technique which will be discussed later.
Possible other numerical strategies concerning this problem are discussed in Kametani et al. (2012) as well as Hurtado and Ortiz (2013).

2.2 Shearing of a Periodic Laminate

In this section, the theory introduced in Sect. 2.1 is applied to an elasto-plastic laminate microstructure exposed to plane strain (Fig. 2.1). Similar problems have been investigated by Cordero et al. (2010), Aslan et al. (2011) as well as Forest and Guéninchault (2013).
The principal objective of the chapter at hand is the presentation of analytical and numerical solutions of the aforementioned simple gradient plasticity model problem. In particular, two promising candidates of the defect energy function W_{g} are investigated concerning their effect on the overall size effects as well as the dislocation pile-up structures building up at impenetrable boundaries. The main motivation of the work is the need for more realistic defect energy functions compared to the quadratic approach. A major deficiency of the latter one represents the necessity of an internal length scale parameter, which usually lacks a physically meaningful interpretation (see Eq. (1.28) with $\omega = 2$). Therefore, energy density functions are considered in the following, the material parameters of which are physically more relevant.
The laminate consists of periodic elastic (hard) and elasto-plastic (soft) layers. The widths of the hard and soft layers are given by h and s, respectively. It is assumed that the plastic layer deforms in single slip with horizontal slip direction $\boldsymbol{d} = \boldsymbol{e}_1$ and vertical slip plane normal $\boldsymbol{n} = \boldsymbol{e}_2$, i.e. $\boldsymbol{H}^{\mathrm{p}} = \gamma \boldsymbol{e}_1 \otimes \boldsymbol{e}_2$. Per definition, the plastic shear strain in the hard phase is set to zero, $\gamma = 0$.
The dislocation density tensor (1.24) can be expressed in terms of the

edge density $\rho_{\vdash} = -\partial_{x_1}\gamma$

$$\boldsymbol{\alpha} = -\rho_{\vdash}\boldsymbol{e}_1 \otimes \boldsymbol{e}_3. \tag{2.22}$$

The quantity $\rho_{\vdash}$ represents the total Burgers vector amount per unit area of edge dislocations. Note, that its unit ($\mu\mathrm{m}^{-1}$) differs from the unit of the total line length per unit volume ρ, given by $\mu\mathrm{m}^{-2}$.

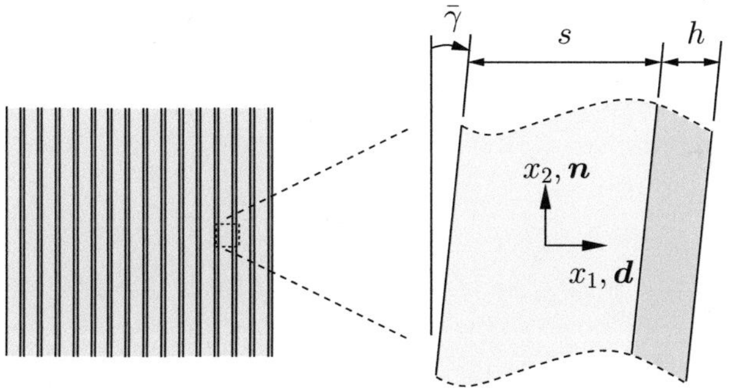

Figure 2.1: Undeformed laminate material and deformed unit cell. The dark elastic phase is hard (h) and the light elasto-plastic phase is soft (s).

Subsequently, the unit cell in Fig. 2.1 (right) is considered. The origin of the coordinate system is located in the center of the soft phase.
The shear deformation is assumed to be given by the following displacement field

$$\boldsymbol{u} = \bar{\gamma}x_2\boldsymbol{e}_1 + \tilde{u}(x_1)\boldsymbol{e}_2. \tag{2.23}$$

The deformation is driven by the macroscopic shear strain $\bar{\gamma}$ (not to be confused with the average of γ)[3].
Since $\tilde{u}(x_1)$ is a periodic fluctuation, the following relations have to be satisfied

$$\int_{-s/2}^{s/2+h} \tilde{u}\,\mathrm{d}x_1 = 0, \qquad \int_{-s/2}^{s/2+h} \tilde{u}'\,\mathrm{d}x_1 = 0. \tag{2.24}$$

[3]Assuming isotropic elastic properties, it can be shown that the fluctuation in e_1-direction vanishes using the linear momentum balance $(2.4)_1$.

Henceforth, $(\bullet)'$ denotes the derivative with respect to x_1. From Eq. (2.23) and the definition of $\boldsymbol{H}^{\mathrm{p}}$, the displacement gradient and elastic strain are found to be

$$\boldsymbol{H} = \bar{\gamma}\boldsymbol{e}_1 \otimes \boldsymbol{e}_2 + \tilde{u}'(x_1)\boldsymbol{e}_2 \otimes \boldsymbol{e}_1 \tag{2.25}$$

and

$$\boldsymbol{\varepsilon}^{\mathrm{e}} = (\bar{\gamma} + \tilde{u}' - \gamma)\mathrm{sym}(\boldsymbol{e}_1 \otimes \boldsymbol{e}_2), \tag{2.26}$$

respectively. The material is taken to be elastically isotropic and homogeneous, for the sake of simplicity. Therefore, Eq. (2.26) implies that all stress components vanish, except for $\sigma_{12} = \sigma_{21}$. From the linear momentum balance $\sigma_{12}'(x_1) = 0$ (see Eq. $(2.4)_1$), it follows that the stress and the elastic strain are homogeneous

$$\varepsilon_{12}^{\mathrm{e}} = \frac{1}{2}(\bar{\gamma} + \tilde{u}' - \gamma) = \frac{\sigma_{12}}{2G} = \text{const.} \tag{2.27}$$

Assuming the defect energy W_{g} to be a function of $\|\boldsymbol{\alpha}\|$, the generalized stress $\boldsymbol{M}$ reads

$$\boldsymbol{M} = \partial_{\alpha} W_{\mathrm{g}} = \partial_{\|\alpha\|} W_{\mathrm{g}} \frac{\boldsymbol{\alpha}}{\|\boldsymbol{\alpha}\|} = -\mathrm{sgn}\,(\rho_{\vdash})\,\partial_{|\rho_{\vdash}|} W_{\mathrm{g}} \boldsymbol{e}_1 \otimes \boldsymbol{e}_3 = M(x_1)\boldsymbol{e}_1 \otimes \boldsymbol{e}_3. \tag{2.28}$$

All other components of $\boldsymbol{M}$ are assumed to vanish[4]. From the balance equation $(2.4)_2$, it follows that

$$s_{12} - M' = 0. \tag{2.29}$$

Throughout this section, the isotropic hardening contribution will be neglected, i.e. $W_{\mathrm{h}} = 0$.

2.2.1 Rank-One Defect Energy

Analytical Solution

For the laminate, the following energy is adopted

$$W_{\mathrm{g}}^{1} = cGb\,\|\boldsymbol{\alpha}\| = cGb\,|\rho_{\vdash}|, \tag{2.30}$$

[4]Note, that this is a quite strong assumption, if $\boldsymbol{\alpha} = \boldsymbol{0}$ and a non-smooth energy are considered. For $\boldsymbol{\alpha} \neq \boldsymbol{0}$, this is a mere consequence of the constitutive equation for $\boldsymbol{M}$.

where c is of order unity (Ortiz and Repetto, 1999). According to Eq. (2.28), the generalized stress M reads

$$\begin{aligned} M = -\tfrac{\rho_\vdash}{|\rho_\vdash|} cGb = -\mathrm{sgn}\,(\rho_\vdash)\, cGb, \quad \text{if } |\rho_\vdash| > 0 \\ |M| \leq cGb, \quad \text{if } |\rho_\vdash| = 0. \end{aligned} \tag{2.31}$$

where the second line follows from Eq. (2.21).

Subsequently, a monotonous shear deformation in the positive direction is prescribed such that the following relations hold in the soft phase

$$\tau^{\text{eff}} \geq \tau_0^{\mathrm{C}}, \quad \dot{\gamma} \geq 0. \tag{2.32}$$

In a first step of the analysis, the flow rule (2.17) is assumed to be given by a linear visco-plastic relation, i.e.

$$\dot{\gamma} = \dot{\gamma}_0 \frac{\tau^{\text{eff}} - \tau_0^{\mathrm{C}}}{\tau^{\mathrm{D}}} \Leftrightarrow \tau^{\text{eff}} = \tau_0^{\mathrm{C}} + \tau^{\mathrm{D}} \frac{\dot{\gamma}}{\dot{\gamma}_0}. \tag{2.33}$$

Note, that the analysis is finally aiming at a rate-independent formulation. At the present stage, the viscous term is still kept to prove that the model leads to a size effect in the model response. Afterwards, viscous terms will be neglected. With Eq. (2.33), the dissipation (2.16) reads

$$\mathcal{D} = \tau^{\text{eff}} \dot{\gamma} = \tau_0^{\mathrm{C}} \dot{\gamma} + \tau^{\mathrm{D}} \frac{\dot{\gamma}^2}{\dot{\gamma}_0}. \tag{2.34}$$

A symmetric slip profile is expected with

$$\gamma'(x_1) \geq 0 \;\; \forall x_1 \in (-s/2, 0), \qquad \gamma'(x_1) \leq 0 \;\; \forall x_1 \in (0, s/2). \tag{2.35}$$

A cuboid-shaped volume $\Delta\mathcal{B}$ is considered, as illustrated in Fig. 2.2. For this volume, the equality of external and internal powers is evaluated in the following.

The power of the external forces reads

$$\mathcal{P}_{\text{ext}}(\Delta\mathcal{B}) = \int_{\partial(\Delta\mathcal{B})} \boldsymbol{t} \cdot \dot{\boldsymbol{u}} \,\mathrm{da} + \int_{\partial(\Delta\mathcal{B})} \boldsymbol{m} \cdot \dot{\boldsymbol{H}}^{\mathrm{p}} \,\mathrm{da}. \tag{2.36}$$

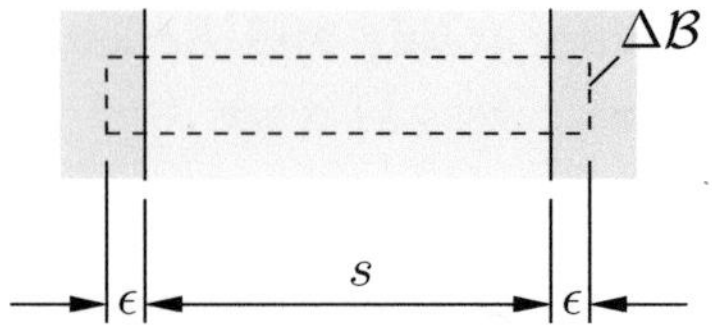

Figure 2.2: Illustration of the integration volume $\Delta\mathcal{B}$.

The second integral vanishes, as shown in App. A.1.
From Gauss' theorem as well as the linear momentum balance $\operatorname{div}(\boldsymbol{\sigma}) = \mathbf{0}$ and the boundary condition $\boldsymbol{t} = \boldsymbol{\sigma}\boldsymbol{n}$ (Eqns. $(2.4)_1$ and $(2.5)_1$), it follows that

$$\mathcal{P}_{\text{ext}}(\Delta\mathcal{B}) = \int\limits_{\partial(\Delta\mathcal{B})} \boldsymbol{t} \cdot \dot{\boldsymbol{u}} \, \mathrm{d}a = \int\limits_{\Delta\mathcal{B}} \boldsymbol{\sigma} \cdot \dot{\boldsymbol{\varepsilon}} \, \mathrm{d}v. \tag{2.37}$$

On the contrary, the power of internal forces is given by

$$\mathcal{P}_{\text{int}}(\Delta\mathcal{B}) = \int\limits_{\Delta\mathcal{B}} p_{\text{int}} \, \mathrm{d}v \overset{(2.16)}{=} \int\limits_{\Delta\mathcal{B}} (\dot{W} + \mathcal{D}) \, \mathrm{d}v. \tag{2.38}$$

Finally, the equality $\mathcal{P}_{\text{ext}}(\Delta\mathcal{B}) = \mathcal{P}_{\text{int}}(\Delta\mathcal{B})$ can be represented as follows (after some rearrangements, see Eq. (2.33) and App. A.2)

$$2cGb\,\dot{\gamma}(0) + (\tau_0^{\text{C}} - \sigma_{12}) \int\limits_{-s/2-\epsilon}^{s/2+\epsilon} \dot{\gamma} \, \mathrm{d}x_1 + \frac{\tau^{\text{D}}}{\dot{\gamma}_0} \int\limits_{-s/2-\epsilon}^{s/2+\epsilon} \dot{\gamma}^2 \, \mathrm{d}x_1 = 0. \tag{2.39}$$

This relation allows to show that there is a size effect on the overall strength of the laminate. In order to prove this, assume that there were no size effect. Then, the solution would be the classical one with $\dot{\gamma}$ =const. in the soft phase. In this case, Eq. (2.39) would reduce to

$$\sigma_{12} = \tau_0^{\text{C}} + \tau^{\text{D}} \frac{\dot{\gamma}}{\dot{\gamma}_0} + \frac{2cGb}{s}, \tag{2.40}$$

after division by $\dot{\gamma}s$. Since the last term scales like $1/s$, the assumption that there is no size effect must be wrong.
The rate-independent limit can be considered by setting $\dot{\gamma}_0 \to \infty$. In this case, the last integral in Eq. (2.39) vanishes. As before, a fully plastic

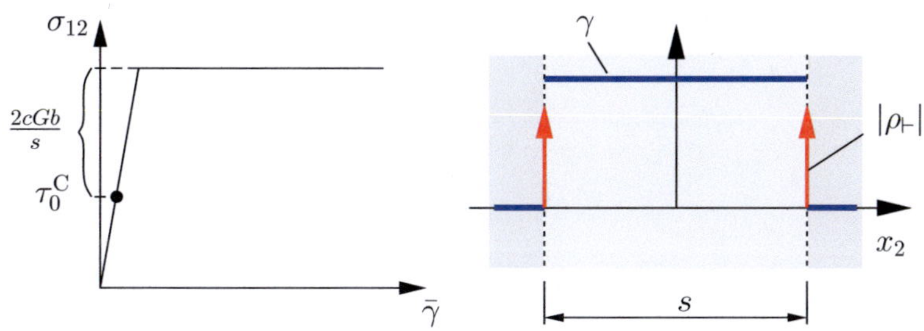

Figure 2.3: Macroscopic shear stress strain curve for the rank-one energy. The increase of the overall yield point scales inversely with the size of the soft phase.

situation is considered, i.e., the yield condition $\tau^{\text{eff}} = \sigma_{12} + s_{12} = \tau_0^{\text{C}}$ is assumed to hold everywhere in the soft phase. Then,

$$M' \overset{(2.29)}{=} s_{12} = -(\sigma_{12} - \tau_0^{\text{C}}) = \text{const.} \tag{2.41}$$

$$\Rightarrow M = -(\sigma_{12} - \tau_0^{\text{C}})x_1, \tag{2.42}$$

where the constant of integration vanishes due to the symmetry requirement $|M(-s/2)| = |M(s/2)|$. Since $|M(x_1)|$ is not constant, the dislocation density $\rho_{\vdash} = -\gamma'$ must vanish (comp. Eq. (2.31)). Therefore, γ =const. and from Eq. (2.39) it follows that

$$\sigma_{12} = \tau_0^{\text{C}} + \frac{2cGb}{s}. \tag{2.43}$$

This equation holds in the plastic regime. Clearly, the application of the rank-one energy increases the macroscopic yield stress by $2cGb/s$, i.e., the increase scales inversely with the size of the soft phase (see Fig. 2.3). The same scaling behavior has been found by Ohno and Okumura (2007) for a spherical grain, also using a rank-one energy. The authors concentrated on the overall mechanical response without having to compute the fields inside of the grains. As illustrated in Fig. 2.3, the dislocations localize in dislocation walls at the elasto-plastic interface.

In the plastic regime, the plastic shear strain follows from Eqns. (2.27) and (2.43) and the constraint $(2.24)_2$

$$\gamma = \frac{s+h}{s}\left(\bar{\gamma} - \frac{\tau_0^{\mathrm{C}} + 2cGb/s}{G}\right). \tag{2.44}$$

In addition to these results, the system behavior can be characterized as follows

- For the material parameters of aluminum ($G = 26.12$ GPa and $b = 0.286$ nm) and $c = 1$, the size effect becomes important when the system size is below $\sim 10\mu$m.
- The plastic shear strain is constant in the bulk, i.e., the dislocations form singularities (walls) at the boundaries.
- The backstress is constant (w.r.t. space) in the bulk. During the first period, it increases and thereby impedes any plastic deformation. Therefore, the overall deformation is purely elastic during this period. At a certain point, the plastic deformation starts and the backstress remains constant afterwards. Its value is given by $2cGb/s$.

2.2.2 Logarithmic Energy

Motivation

This section investigates the following defect energy

$$W_{\mathrm{g}}^{\ln} = c_0 \|\boldsymbol{\alpha}\| \ln \frac{\|\boldsymbol{\alpha}\|}{\alpha_0}, \tag{2.45}$$

with the constant c_0 as defined in Eq. (2.13). The energy is motivated by the statistical theory of dislocations by Groma et al. (2003). The authors derive a backstress term which involves the second gradient of slip as postulated by Aifantis (1987). However, their theory involves an internal length scale which is given by $1/\sqrt{\rho}$, where ρ denotes the total dislocation density.

In pure metals, the geometrical characteristics of the microstructure are essentially determined by the dislocation arrangement. This is a strong

argument for a (variable) internal length scale, which is determined by the available dislocation field variables (instead of a constant length scale parameter, see also Forest and Sedláček (2003) where this dependency is derived from a dislocation line tension model).
It is demonstrated subsequently, that the approach (2.45) leads to a backstress which is similar to that of Groma et al. (2003). However, it should be mentioned that this energy is neither convex nor smooth with respect to the dislocation density tensor (a regularization will be discussed at a later stage).
For the laminate problem, the generalized stress M reads (see Eq. (2.28))

$$M = -\mathrm{sgn}\,(\rho_{\vdash})\; c_0 \left(\ln \frac{|\rho_{\vdash}|}{\alpha_0} + 1\right). \tag{2.46}$$

In this section, rate-independent plasticity will be considered based on the yield criterion

$$f = |\tau^{\mathrm{eff}}| - \tau_0^{\mathrm{C}} \leq 0. \tag{2.47}$$

Here, the effective stress reads

$$\tau^{\mathrm{eff}} = (\boldsymbol{\sigma} + \boldsymbol{s}) \cdot (\boldsymbol{d} \otimes \boldsymbol{n}) = \sigma_{12} + s_{12} \overset{(2.29)}{=} \sigma_{12} + M'. \tag{2.48}$$

With Eq. (2.46) and $M' = (\partial_{\rho_{\vdash}} M)\,(\partial_{x_1} \rho_{\vdash})$, it follows that

$$\tau^{\mathrm{eff}} = \tau - \frac{c_0}{|\rho_{\vdash}|} \partial_{x_1} \rho_{\vdash} = \tau + \frac{G c_1}{2\pi(1-\upsilon)} \frac{b}{|\rho_{\vdash}|} \partial_{x_1}^2 \gamma. \tag{2.49}$$

Here, the second term can be interpreted as a backstress. Note that the backstress involves no internal length scale parameter. Instead, the internal length scale, $\sqrt{b/\rho_{\vdash}}$, is determined by the dislocation microstructure. In contrast to the backstress of Groma et al. (2003), the internal length scale is determined by the GND-density $\rho_{\vdash}$ instead of the total density ρ. Hence, the influence of statistically stored dislocations (SSDs) is ignored. This question will be addressed at a later point. For the moment, GND-dominated problems will be focused on. Therefore, a homogeneous initial GND-density $|\rho_{\vdash}| = \alpha_0$ will be assumed to be given. In addition, it is assumed that the SSD-density is equal or less than α_0.

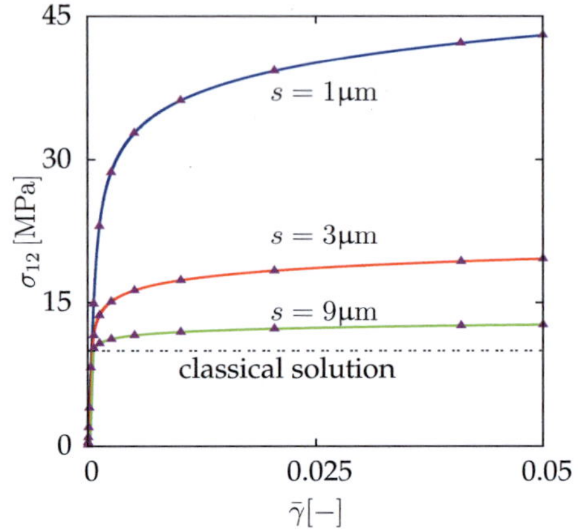

Figure 2.4: Macroscopic stress-strain diagram for three different sizes. Analytical (lines) and regularized, numerical (triangles) solution for the logarithmic potential.

Analytical Solution

The soft phase is assumed to be under plastic loading, with $\tau^{\text{eff}} = \tau_0^{\text{C}}$ in the soft phase. In this case,

$$M' \overset{(2.29)}{=} s_{12} = -(\sigma_{12} - \tau_0^{\text{C}}) = \text{const.} \tag{2.50}$$

$$\Rightarrow M = -(\sigma_{12} - \tau_0^{\text{C}})x_1, \tag{2.51}$$

where, again, the constant of integration vanishes due to the symmetry requirement $|M(-s/2)| = |M(s/2)|$. The plastic slip γ can be derived from the equality of Eqns. (2.46) and (2.51), which yields a differential equation for γ. The solution reads

$$\gamma = \frac{\alpha_0 L_\tau}{e}\left(\exp\left(\frac{s}{2L_\tau}\right) - \exp\left(-a\frac{x_1}{L_\tau}\right)\right) \quad \text{with} \quad L_\tau = \frac{c_0}{\sigma_{12} - \tau_0^{\text{C}}}, \tag{2.52}$$

where the matching conditions $\gamma(-s/2) = \gamma(s/2) = 0$ have been exploited and where $e = \exp(1)$. The variable a is defined by $a = \text{sgn}(\gamma')$ and is assumed positive in $(-s/2, 0)$ and negative in $(0, s/2)$ [5].

[5] It is noteworthy, that the solution depends on α_0, although this constant does neither appear in the field equation $\tau_0^{\text{C}} = \sigma_{12} + c_0\partial^2_{x_1}\gamma/|\partial_{x_1}\gamma|$ nor in the matching condition $\gamma(\pm s/2) = 0$.

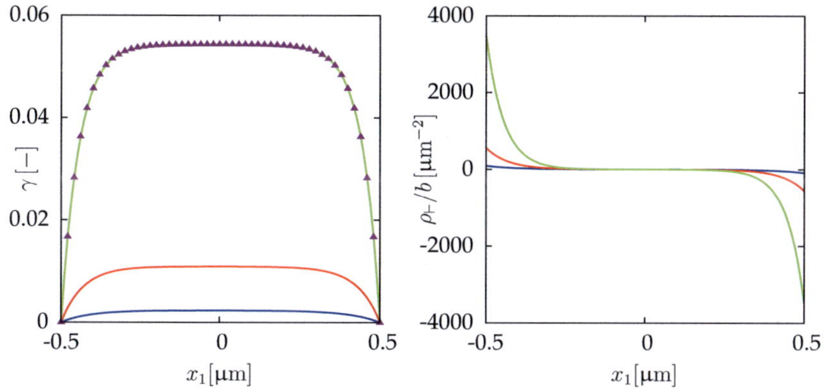

Figure 2.5: Plastic slip γ and dislocation density $\rho_{\vdash}/b$ for the smallest size ($s = 1\mu\mathrm{m}$), according to the logarithmic model. Left: Plastic slip at 0.3, 1.1 and 5 % macroscopic strain (triangles: regularized, numerical solution). Right: corresponding GND-densities.

From Eqns. $(2.24)_2$ and (2.27) the macroscopic stress strain relation follows

$$\bar{\gamma} = \frac{\alpha_0 L_\tau}{e(s+h)} \left(\exp\left(\frac{s}{2L_\tau} \right)(s - 2L_\tau) + 2L_\tau \right) + \frac{\sigma_{12}}{G}. \tag{2.53}$$

The solution is evaluated for the following material parameters

E [GPa]	v	τ_0^{C} [MPa]	b [nm]	c_1	α_0/b [μm^{-2}]
70	0.34	10	0.286	1	1

Here, E and v denote Young's modulus and Poisson's ratio. A very thin hard phase with negligible width h is considered ($h/s = 10^{-6}$ for the analytical solution).

The macroscopic stress-strain curve (2.53) is illustrated in Fig. 2.4. A clear size effect is visible. Apparently, mainly the overall yield stress is affected. The hardening shows less size dependence. It is remarkable that the model provides a size–dependent yield stress and non–linear kinematic hardening.

The microscopic plastic shear strain γ and dislocation density are shown in Fig. 2.5. Dislocation pile-ups are observed at the boundaries of the soft phase.

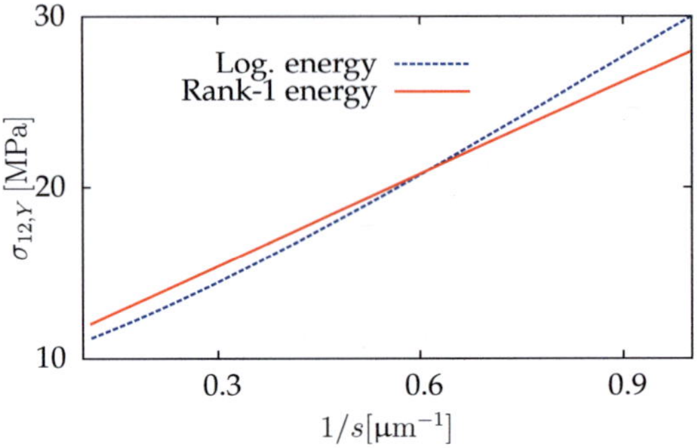

Figure 2.6: Scaling-comparison of the two energies for $c = 1.2$. The shear stress values of the logarithmic energy correspond to 0.2% plastic shear strain.

Since there is no distinct yield stress, the evaluation of the scaling behavior is based on the offset yield stress at 0.2% plastic strain. Fig. 2.6 shows the offset yield stress as a function of the inverse of the size $1/s$. For comparison, the results obtained from the rank-one energy are illustrated, in addition. For $c = 1.2$, the scaling behavior of both energies is similar in the considered range. The cyclic loading curves are represented in Fig. 2.7 for $s = 3\mu m$. The influence of the backstress is clearly observable. The curves in Fig. 2.7 have been obtained numerically. Therefore, both models are regularized as explained in the subsequent sections.

The type of non–linear kinematic hardening observed for both models corresponds to Asaro's type KIII model, corresponding to a *first in / last out* sequence of dislocation motion (Asaro (1975)). It is considered by Asaro as the most perfect form of recovery of plastic memory. Such stress–strain loops display inflection points that are observed in some materials, see Asaro (1975) for a Nimonic alloy, but such observations have also been made in several Nickel based superalloys. It is usually attributed to substructural recovery on the microscale, for instance pile–up formation and destruction at γ' precipitates. In the present simple single crystal model, it is the only active hardening mechanism induced

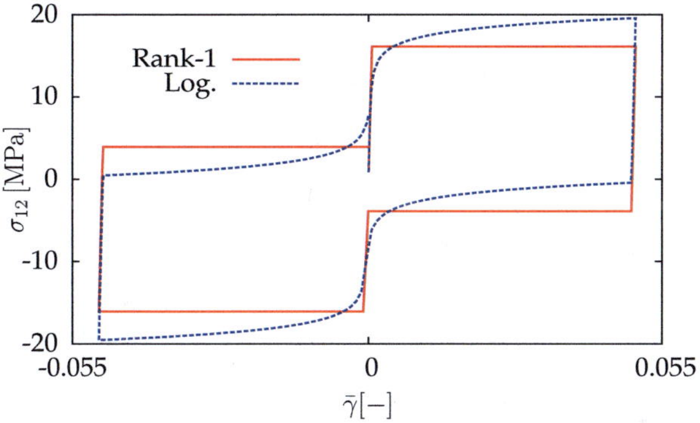

Figure 2.7: Cyclic loading for $s = 3\mu\text{m}$.

by the strain gradient plasticity effects and the presence of the hard phase in the laminate. It represents an accurate continuum description of piling-up and unpiling-up dislocation phenomena.

Regularization of the Logarithmic Energy

The length scale $\sqrt{b/\rho_\vdash}$ of the backstress in Eq. (2.49) is determined by the GNDs. This was the main motivation of the logarithmic energy (Eq. (2.45)). In the following, the theory is extended to problems which are not fully GND-dominated. For that purpose, the following regularization is introduced (see Fig. 2.8)

$$W_{\mathrm{g}} = \begin{cases} \frac{1}{2}\frac{c_0}{b} l^2 \|\boldsymbol{\alpha}\|^2, & \|\boldsymbol{\alpha}\| < \alpha_{\mathrm{L}} \\ c_0 \|\boldsymbol{\alpha}\| \ln \frac{\|\boldsymbol{\alpha}\|}{\alpha_0} + W_0, & \text{else.} \end{cases} \tag{2.54}$$

In the region of small GND-densities, the energy is replaced by a quadratic potential. The internal length scale l, the transition density α_{L} and the offset energy W_0 are chosen such that W_{g}, $\partial_{\|\alpha\|} W_{\mathrm{g}}$ and $\partial^2_{\|\alpha\|} W_{\mathrm{g}}$ are continuous at the transition point $\|\boldsymbol{\alpha}\| = \alpha_{\mathrm{L}}$. As a result

$$\alpha_{\mathrm{L}} = \alpha_0, \quad l^2 = \frac{b}{\alpha_0}, \quad W_0 = \frac{c_0 \alpha_0}{2}. \tag{2.55}$$

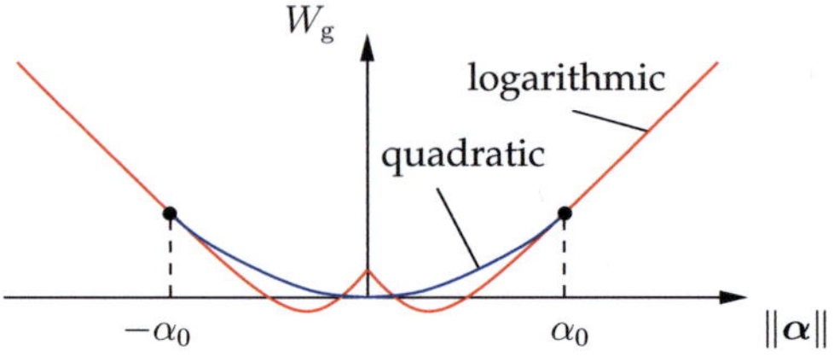

Figure 2.8: Regularization of the logarithmic energy.

The regularized energy (2.54) is convex, normalized and twice differentiable. The backstress for the laminate problem reads

$$x = \begin{cases} -\frac{c_0}{\alpha_0}\partial^2_{x_1}\gamma, & |\rho_\vdash| < \alpha_0, \\ -\frac{c_0}{|\rho_\vdash|}\partial^2_{x_1}\gamma, & \text{else.} \end{cases} \tag{2.56}$$

The solution based on the regularized model has been obtained numerically by finite elements. Figs. 2.4 and 2.5 compare the analytical and numerical solution.

Interpretation of the Regularization

If the dislocation microstructure is GND-dominated, the internal length scale of the laminate model is supposed to be $\sqrt{b/\rho_\vdash}$, i.e., determined by the GND-density. Therefore, the threshold α_0 in Eq. (2.56) should be chosen such that it clearly indicates whether the microstructure at a given point is GND- or SSD-dominated. If the SSD-density is nearly homogeneous, a reasonable choice of α_0 is the SSD-density itself. If the SSD-density is not constant, α_0 might be interpreted as a characteristic SSD-density. In this case, the logarithmic energy is applied, if $|\rho_\vdash| > \alpha_0$, i.e., if the problem is GND-dominated.

However, the quadratic energy is applied if $|\rho_\vdash| < \alpha_0$, i.e., when the SSDs dominate. As an interesting feature, the internal length scale $\sqrt{b/\alpha_0}$ of the backstress (2.56) is then determined by the SSD-density.

In both cases, the backstress (2.56) is approximately given by

$$x \approx -\frac{c_0/b}{\rho}\partial^2_{x_1}\gamma. \tag{2.57}$$

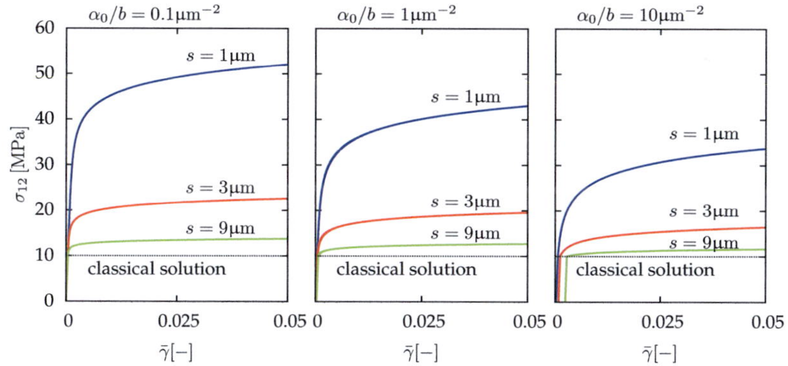

Figure 2.9: Influence of α_0 on the overall size effect, according to the regularized logarithmic model.

This backstress coincides with Groma's representation. In this sense, the energy approach (2.54) reduces to a gradient plasticity approximation of Groma's theory. Eq. (2.57) is valid except of the case $\rho_{\mathrm{SSD}} \approx \rho_{\mathrm{GND}}$. In this case, the backstress is overestimated by a factor of two.

Fig. 2.9 shows the influence of the parameter α_0 on the size effect. The parameter α_0 has been varied within three orders of magnitude ($\alpha_0/b = 0.1$, 1 and 10 μm^{-2}). Obviously, a variation of α_0 within orders of magnitude is indeed necessary in order to significantly influence the results. Hence, the model sensitivity with respect to changes of α_0 is smaller than might have been expected.

One might hope that a rough estimate of the SSD-density could be sufficient to achieve a reasonable guess of the parameter α_0. In particular, the overestimation of the backstress in the case $\rho_{\mathrm{GND}} = \rho_{\mathrm{SSD}}$ by a factor of two has less consequences than expected, since a multiplication of α_0 by two has a minor influence on the results.

Here, the sensitivity of σ_{12} with respect to changes of α_0 has been investigated for a specific set of material parameters. A more general sensitivity analysis, as given in App. A.3, leads to similar results.

2.3 Numerical Solution Strategy

The numerical solution strategy is formulated for the general three-dimensional case. For the laminate problem, the implementation can be simplified in a straightforward way.

2.3.1 Incremental Potential

It is convenient to introduce slip parameters $\hat{\lambda} = (\lambda_1, \lambda_2, ..., \lambda_{2N})^{\mathrm{T}}$ for the numerical implementation, where N is the number of slip systems. Two slip parameters per slip system are introduced which account for positive and negative slip increments separately. These are associated to the positive and negative slip directions $\boldsymbol{d}_\alpha$ and $-\boldsymbol{d}_\alpha$, i.e., the total slip γ_α of a given slip system is represented by the difference of the associated two slip parameters. As a consequence, the flow rule (2.17) is replaced by

$$\dot{\lambda}_\alpha = \dot{\gamma}_0 \left\langle \frac{\tau_\alpha^{\mathrm{eff}} - (\tau_0^{\mathrm{C}} + q)}{\tau^{\mathrm{D}}} \right\rangle^p , \tag{2.58}$$

where, for simplicity, it is assumed that γ_{eq} is the only history variable, i.e. $\hat{\mu} = (\gamma_{\mathrm{eq}})$.

The numerical solution is based on the implicit Euler scheme. This means, that the time is discretized into steps. Quantities of the preceding time step are marked by an index "n". For convenience, the index "n + 1" is dropped. Increments are marked by the symbol "Δ". Subsequently, it will be discussed that the overall problem can be reformulated as the stationarity conditions of the following potential

$$\Pi = \Pi_{\mathrm{W}} + \Pi_{\mathrm{L}} + \Pi_{\mathcal{D}} + \Pi_{\mathrm{ext}}, \tag{2.59}$$

with a contribution from the free energy

$$\Pi_{\mathrm{W}} = \int_{\mathcal{B}} \left(W_{\mathrm{e}}(\mathrm{sym}(\nabla \boldsymbol{u}), \boldsymbol{\varepsilon}^{\mathrm{p}}(\hat{\lambda})) + W_{\mathrm{g}}(\boldsymbol{\alpha}) + W_{\mathrm{h}}(\gamma_{\mathrm{eq}}(\hat{\lambda})) \right) \mathrm{d}v, \tag{2.60}$$

a Lagrange-contribution

$$\Pi_{\mathrm{L}} = \int_{\mathcal{B}} \left(\boldsymbol{s} \cdot \left(\boldsymbol{H}^{\mathrm{p}} - \sum_\alpha \lambda_\alpha \boldsymbol{d}_\alpha \otimes \boldsymbol{n}_\alpha \right) + \boldsymbol{M} \cdot \left(\mathrm{curl}^{\mathrm{T}}(\boldsymbol{H}^{\mathrm{p}}) - \boldsymbol{\alpha} \right) \right) \mathrm{d}v, \tag{2.61}$$

a dissipative contribution

$$\Pi_{\mathcal{D}} = \int_{\mathcal{B}} \Delta t \Bigg(\sum_{\alpha} \tau_{\alpha}^{\mathrm{d}} \frac{\Delta \lambda_{\alpha}}{\Delta t} - \frac{\dot{\gamma}_0 \, \tau^{\mathrm{D}}}{p+1} \left\langle \frac{\tau_{\alpha}^{\mathrm{d}} - \tau_0^{\mathrm{C}}}{\tau^{\mathrm{D}}} \right\rangle^{p+1} \Bigg) \,\mathrm{d}v \tag{2.62}$$

and an external force contribution

$$\Pi_{\mathrm{ext}} = - \int_{\partial \mathcal{B}_t} \bar{\boldsymbol{t}} \cdot \boldsymbol{u} \,\mathrm{d}a - \int_{\partial \mathcal{B}_m} \bar{\boldsymbol{m}} \cdot \boldsymbol{H}^{\mathrm{p}} \,\mathrm{d}a. \tag{2.63}$$

The stationarity conditions associated to the potential Π with respect to the (a priori independent) arguments $\{\boldsymbol{u}, \boldsymbol{H}^{\mathrm{p}}, \hat{\lambda}, \boldsymbol{s}, \boldsymbol{M}, \boldsymbol{\alpha}, \hat{\tau}^{\mathrm{d}}\}$ yield the system equations. Here $\hat{\tau}^{\mathrm{d}} = (\tau_1^{\mathrm{d}}, \tau_2^{\mathrm{d}}, ..., \tau_{2N}^{\mathrm{d}})^{\mathrm{T}}$ can be interpreted as dissipative shear stresses. The principle of virtual power (Eq. (2.3)) is recovered by the conditions $\delta_{\boldsymbol{u}} \Pi = 0$ and $\delta_{\boldsymbol{H}^{\mathrm{p}}} \Pi = 0$. Note, that the relations $\boldsymbol{\sigma} = \partial_{\boldsymbol{\varepsilon}} W_{\mathrm{e}}$ with $\boldsymbol{\varepsilon} = \mathrm{sym}(\nabla \boldsymbol{u})$ are implicitly accounted for. Equations (1.8) and (2.8) are obtained by $\delta_{\boldsymbol{M}} \Pi = 0$ and $\delta_{\boldsymbol{s}} \Pi = 0$, respectively. These relations illustrate the interpretation of the stresses $\boldsymbol{M}$ and $\boldsymbol{s}$ as Lagrange multipliers which enforce the kinematic relations (1.8) and (2.8).

Evaluation of the condition $\delta_{\boldsymbol{\alpha}} \Pi = 0$ yields the relation $\boldsymbol{M} = \partial_{\boldsymbol{\alpha}} W_{\mathrm{g}}$. Finally, the conditions $\delta_{\lambda_\alpha} \Pi = 0$ and $\delta_{\tau_\alpha^{\mathrm{d}}} \Pi = 0$ yield in combination the implicit Euler scheme associated to the flow rule

$$\frac{\Delta \lambda_{\alpha}}{\Delta t} = \dot{\gamma}_0 \left\langle \frac{\tau_{\alpha}^{\mathrm{eff}} - (\tau_0^{\mathrm{C}} + q)}{\tau^{\mathrm{D}}} \right\rangle^{p}, \tag{2.64}$$

with $\tau_{\alpha}^{\mathrm{eff}} = (\partial_{\boldsymbol{\varepsilon}} W + \boldsymbol{s}) \cdot (\boldsymbol{d}_{\alpha} \otimes \boldsymbol{n}_{\alpha})$ and $\boldsymbol{\varepsilon} = \mathrm{sym}(\nabla \boldsymbol{u})$.

2.3.2 Augmented Lagrange Multiplier and Penalty Method

The augmented Lagrange Multiplier Method is based on the replacement of the Lagrange multipliers $\boldsymbol{M}$ and $\boldsymbol{s}$ in Eq. (2.61) by

$$\boldsymbol{M} = \boldsymbol{M}^{\mathrm{old}} + H_{\chi}^{\mathrm{M}} (\mathrm{curl}^{\mathrm{T}}(\boldsymbol{H}^{\mathrm{p}}) - \boldsymbol{\alpha}), \tag{2.65}$$

$$\boldsymbol{s} = \boldsymbol{s}^{\mathrm{old}} + H_{\chi}^{\mathrm{s}} \Big(\boldsymbol{H}^{\mathrm{p}} - \sum_{\alpha} \lambda_{\alpha} \boldsymbol{d}_{\alpha} \otimes \boldsymbol{n}_{\alpha} \Big). \tag{2.66}$$

Here, H_χ^{M} and H_χ^{s} are penalty parameters and $\boldsymbol{M}^{\mathrm{old}}$ as well as $\boldsymbol{s}^{\mathrm{old}}$ are approximations of the actual Lagrange multipliers. As a consequence of the replacements (2.65) and (2.66), the set of arguments of the potential Π (Eq. (2.59)) reduces to $\{\boldsymbol{u}, \boldsymbol{H}^{\mathrm{p}}, \hat{\lambda}, \boldsymbol{\alpha}, \hat{\tau}^{\mathrm{d}}\}$. Once a converged solution has been obtained (i.e. the stationarity conditions are satisfied), the following update is effectuated

$$\boldsymbol{M}^{\mathrm{old}} \leftarrow \boldsymbol{M}^{\mathrm{old}} + H_\chi^{\mathrm{M}}(\mathrm{curl}^{\mathrm{T}}(\boldsymbol{H}^{\mathrm{p}}) - \boldsymbol{\alpha}), \tag{2.67}$$

$$\boldsymbol{s}^{\mathrm{old}} \leftarrow \boldsymbol{s}^{\mathrm{old}} + H_\chi^{\mathrm{s}}\left(\boldsymbol{H}^{\mathrm{p}} - \sum_\alpha \lambda_\alpha \boldsymbol{d}_\alpha \otimes \boldsymbol{n}_\alpha\right) \tag{2.68}$$

and the computation of the solution is repeated with an updated set of Lagrange multipliers yielding an improved approximation of Eqns. (2.8) and (1.8). The overall procedure is reiterated until Eqns. (2.8) and (1.8) are satisfied up to a certain tolerance.

If H_χ^{M} and H_χ^{s} are very large, the solution might be sufficiently accurate after one iteration already. In this case, the updates (2.67) and (2.68) are dispensable and can be omitted. Then, the scheme represents a penalty method (which has been used to obtain the numerical results in the work at hand with $H_\chi^{\mathrm{M}} = H_\chi^{\mathrm{s}} = 10^6$ MPa). Note that, in contrast to the original model (2.18), $\boldsymbol{M}$ can be computed even if $\boldsymbol{\alpha} = \boldsymbol{0}$. The close connection to classical elasto-plasticity will become obvious during the subsequent discussion of the local algorithm 2.

2.3.3 Local Algorithms

The principle of virtual power (2.3) is discretized by the finite element method. At the integration points, the following two algorithms allow the determination of the stresses.

Algorithm 1. Compute the stresses $\{\boldsymbol{\sigma}, \boldsymbol{s}, q\}$ by solving the nonlinear system of equations

$$\begin{aligned}
\boldsymbol{0} &= -\mathbb{S}[\boldsymbol{\sigma}] + \boldsymbol{\varepsilon} - \boldsymbol{\varepsilon}_{\mathrm{n}}^{\mathrm{p}} - \sum_\alpha \Delta t \dot{\gamma}_0 \left\langle \frac{(\boldsymbol{\sigma}+\boldsymbol{s})\cdot(\boldsymbol{d}_\alpha \otimes \boldsymbol{n}_\alpha) - q - \tau_0^{\mathrm{C}}}{\tau^{\mathrm{D}}} \right\rangle^p \mathrm{sym}(\boldsymbol{d}_\alpha \otimes \boldsymbol{n}_\alpha) \\
\frac{\boldsymbol{s}}{H_\chi^{\mathrm{s}}} &= \frac{\boldsymbol{s}^{\mathrm{old}}}{H_\chi^{\mathrm{s}}} + \boldsymbol{H}^{\mathrm{p}} - \sum_\alpha \left(\lambda_{\alpha,n} + \Delta t \dot{\gamma}_0 \left\langle \frac{(\boldsymbol{\sigma}+\boldsymbol{s})\cdot(\boldsymbol{d}_\alpha \otimes \boldsymbol{n}_\alpha) - q - \tau_0^{\mathrm{C}}}{\tau^{\mathrm{D}}} \right\rangle^p\right) \boldsymbol{d}_\alpha \otimes \boldsymbol{n}_\alpha \\
0 &= q - \partial_{\gamma_{\mathrm{eq}}} W_{\mathrm{h}}.
\end{aligned}$$

Here, $\mathbb{S} = \mathbb{C}^{-1}$ denotes the compliance tensor. The linearization of this system of equations yields the consistent linearizations (tangent operators) of the stresses $\{\boldsymbol{\sigma}, \boldsymbol{s}, q\}$ (for details see, e.g., Wulfinghoff and Böhlke, 2013, for a similar problem).

Algorithm 2. The following non-smooth energy is considered

$$W_{\rm g} = cGb\|\boldsymbol{\alpha}\| + W_{\rm g}^{\rm s}(\|\boldsymbol{\alpha}\|), \tag{2.69}$$

where $W_{\rm g}^{\rm s} = W_{\rm g}^{\rm s}(\|\boldsymbol{\alpha}\|)$ is assumed to be convex and smooth. For simplicity, it is assumed that $W_{\rm g}^{\rm s} = El_{\rm s}^2\|\boldsymbol{\alpha}\|^2/2$. Then the following algorithm can be applied.
Trial step: Set $\tilde{\boldsymbol{\alpha}} = \mathrm{curl}^{\rm T}(\boldsymbol{H}^{\rm p})$

$$\boldsymbol{\alpha}^{\rm Tr} = \boldsymbol{0}, \quad \boldsymbol{M}^{\rm Tr} = \boldsymbol{M}^{\rm old} + H_\chi^{\rm M}(\tilde{\boldsymbol{\alpha}} - \boldsymbol{\alpha}^{\rm Tr}), \quad \varphi_{\rm g}^{1,{\rm Tr}} = \|\boldsymbol{M}^{\rm Tr}\| - cGb. \tag{2.70}$$

If $\varphi_{\rm g}^{1,{\rm Tr}} \leq 0$

$$\boldsymbol{\alpha} = \boldsymbol{\alpha}^{\rm Tr} = \boldsymbol{0}, \quad \boldsymbol{M} = \boldsymbol{M}^{\rm Tr} \tag{2.71}$$

else set

$$\boldsymbol{N}^{\rm Tr} = \frac{\boldsymbol{M}^{\rm Tr}}{\|\boldsymbol{M}^{\rm Tr}\|}, \quad \boldsymbol{\alpha} = \frac{\varphi^{\rm Tr}}{H_\chi^{\rm M} + El_{\rm s}^2}\boldsymbol{N}^{\rm Tr}, \quad \boldsymbol{M} = \boldsymbol{M}^{\rm Tr} - H_\chi^{\rm M}\boldsymbol{\alpha}. \tag{2.72}$$

For the FE-implementation, the linearization of $\boldsymbol{M}$ w.r.t. $\tilde{\boldsymbol{\alpha}}$ can be obtained by the linearization of the algorithm.
Note the formal similarity of algorithm 2 and classical radial-return algorithms (e.g. Simo and Hughes, 1998).
In the work at hand, a linear-viscous model ($p = 1$) with very small viscosity ($\dot{\gamma}_0 = 10^3{\rm s}^{-1}$, $\tau^{\rm D} = 1$MPa) has been applied. The total simulation time was 1s. The penalty approximation (2.67) (with $\boldsymbol{M}^{\rm old} = \boldsymbol{0}$) has only been used for the rank-one potential. This is not necessary for the regularized logarithmic energy (2.54), which is smooth. The problem has been solved using 50 one-dimensional linear elements. The results correspond to the Figs. 2.4, 2.5 and 2.7 in Sect. 2.2.

Chapter 3

Approximation of a Model Material With Localized Slip Bands by a Gradient Plasticity Theory

3.1 Introduction

Several crystal gradient plasticity theories are based on the dislocation density tensor (Eq. (1.22)) or on scalar GND-densities (Eq. (1.24)). These are reasonable kinematical measures, since they describe actual features of the underlying dislocation microstructure as discussed in Chap. 1. However, the associated stresses (like the higher order stresses ξ_α, Eq. (1.26), or $\boldsymbol{M}$, Eq. (2.15)) are generally less intuitively understandable than conventional stresses, like the resolved shear stress or the Cauchy stress tensor. Therefore, the behavior of gradient plasticity models is usually neither easy to predict nor to understand. However, in many cases it is necessary to have a clear idea of the model behavior in mind, for example to set proper boundary conditions.

In order to improve the understanding of gradient plasticity theories, an idealized crystal plasticity model is introduced. The microstructure of this model is simple, such that its mechanical behavior can be understood and predicted intuitively. It is then shown, that the material behavior of this model is in some situations approximately described by standard gradient plasticity equations. For this simple material, most usually abstract quantities – like the internal length scale, microstresses or microtractions – have a clear and intuitive character. Having in mind that this simplified material model does not represent the real single crystal microstructure, it still can be quite helpful to improve the overall understanding of phenomenological gradient theories and get an insight into the following aspects of gradient plasticity

- the idea of an internal length scale
- the mechanical character of boundary microtractions
- the interaction of these microtractions with the bulk
- a possible interpretation of microstresses and their long-range character

The presented idealized material model is formulated based on the following assumptions

- a single slip situation is considered
- the plastic slip is perfectly localized in discrete slip bands with constant distance h, which takes the role of the internal length scale
- the layers in between the slip bands deform perfectly elastically
- except for the localization of the plastic slips, the deformation is assumed smooth, in particular the characteristic wavelengths of the displacement field are sufficiently large compared to the slip band spacing

Although idealized, the material model is not purely artificial, since there are indeed experimental examples where the plastic deformation of single crystals is to a high extend localized in slip bands. However, this chapter does not aim at a discussion concerning the situations in which the aforementioned assumptions might represent real single crystal behavior. Instead, the model is supposed to serve as an illustrative example for a gradient plasticity theory which is ought to give some intuitive ideas concerning the meaning of terms like microstresses and micro-force balances.

In addition, the considerations might also be interesting for an application to layered composites, if a relative motion (e.g. due to creep) between the layers is possible.

Originally, a similar but more substantial theory has been proposed by Mesarovic et al. (2010) (see also Mesarovic, 2005; Yassar et al., 2007; Baskaran et al., 2010; Mesarovic, 2010). Their theory is applicable to multislip situations and comprises advanced grain boundary models. The theory outlined in this chapter can be considered as a similar but simplified version thereof. It is emphasized that the principal goal of the chapter at hand is the formulation of a gradient plasticity theory which is first of all illustrative.

3.2 Slip Band Size Effects

In this section, two simple examples demonstrate the accuracy limits of classical crystal plasticity, as outlined in Sect. 1.4, when the plastic deformation is not fully continuous but localized in the form of slip bands. Therefore, a bending and a torsion example are discussed. It is shown, that the continuum mechanical crystal plasticity model might be applicable in situations where the characteristic wavelengths of the deformation are much larger than the slip band spacing. However, the continuum model becomes inaccurate when the aforementioned wavelengths approach the order of magnitude of the slip band spacing.

3.2.1 Plastic Bending

Continuous Plastic Deformation

Let the displacement field of the cube shaped single crystal with dimensions L_{c} in Fig. 3.1 be given by

$$\boldsymbol{u}(\boldsymbol{x}) = u(\boldsymbol{x})\boldsymbol{e}_x + v(\boldsymbol{x})\boldsymbol{e}_y + w(\boldsymbol{x})\boldsymbol{e}_z. \tag{3.1}$$

Let the displacement field and the plastic deformation be *given* (subsequently, it will be shown that the body is in equilibrium).
A simple bending deformation is considered by setting

$$w = w(x) = \frac{\gamma_0}{2L_{\mathrm{c}}}x^2 \tag{3.2}$$

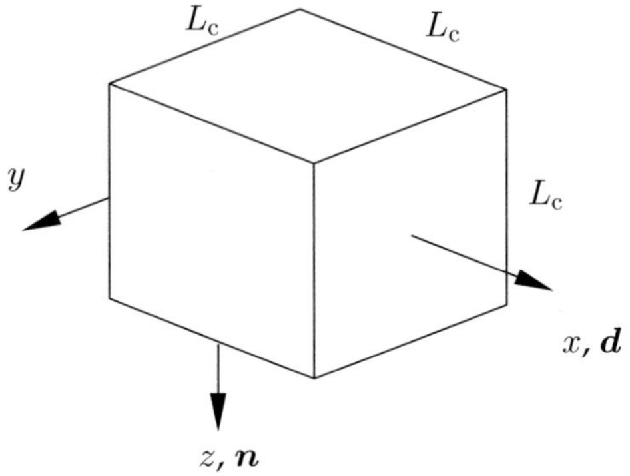

Figure 3.1: Cube suspended to bending.

and $u = v = 0$, respectively. Here, γ_0 is a constant.
The displacement gradient reads

$$\boldsymbol{H} = \nabla \boldsymbol{u} = \gamma_0 \frac{x}{L_c} \boldsymbol{e}_z \otimes \boldsymbol{e}_x. \tag{3.3}$$

A single slip situation is considered with

$$\gamma_1 = \gamma(x) = \gamma_0 \frac{x}{L_c} \tag{3.4}$$

associated to the slip system with direction $\boldsymbol{d}_1 = \boldsymbol{d} = \boldsymbol{e}_x$ and slip plane normal $\boldsymbol{n}_1 = \boldsymbol{n} = \boldsymbol{e}_z$. The plastic part of the displacement gradient reads

$$\boldsymbol{H}^{\mathrm{p}} = \gamma(x) \boldsymbol{e}_x \otimes \boldsymbol{e}_z. \tag{3.5}$$

Assuming additivity of elastic and plastic strains, the elastic strain tensor is given by

$$\boldsymbol{\varepsilon}^{\mathrm{e}} = \boldsymbol{\varepsilon} - \boldsymbol{\varepsilon}^{\mathrm{p}} = \mathrm{sym}(\boldsymbol{H}) - \mathrm{sym}(\boldsymbol{H}^{\mathrm{p}}) = \boldsymbol{0}. \tag{3.6}$$

If Hooke's law is applied, the stress tensor vanishes

$$\boldsymbol{\sigma} = \mathbb{C}[\boldsymbol{\varepsilon}^{\mathrm{e}}] = \boldsymbol{0}, \tag{3.7}$$

with the elastic stiffness tensor $\mathbb{C}$. Accordingly, the elastic strain energy also vanishes

$$W_{\mathrm{e}} = \frac{1}{2} \boldsymbol{\varepsilon}^{\mathrm{e}} \cdot \mathbb{C}[\boldsymbol{\varepsilon}^{\mathrm{e}}] = 0. \tag{3.8}$$

This means that the bending deformation is purely plastic.

Localized Plastic Deformation

Now the same deformation is applied to a body assuming the plastic deformation strictly localized in planes which separate isotropic, purely elastic layers of constant thickness h, see Fig. 3.2. The total number of layers is given by $N_\mathrm{c} = L_\mathrm{c}/h$. The layers are assumed to behave like simple beams (i.e. they are described by a one-dimensional generalized continuum theory). The Bernoulli hypothesis is applied, i.e., the cross sections are assumed to be perpendicular to the center line (connecting the centroids of all cross sections). This approach is a strong simplification since it completely ignores the influence of lateral contractions (as it is usually done in bending theory). Strictly speaking, this approach is only correct for $v \to 0$ (v is Poisson's ratio). However, it is applied here since it is well known and at this point, this level of accuracy is sufficient. Let the deflection of the individual beams be given by Eq. (3.2). The axial moment of inertial of an individual layer with index i reads $I = L_\mathrm{c}h^3/12$ and its strain energy can be computed by

$$W_i = \frac{1}{2}\int_0^{L_\mathrm{c}} EI(w''(x))^2\,\mathrm{d}x = \frac{1}{24}L_\mathrm{c}^2 Eh^3(\gamma'(x))^2, \tag{3.9}$$

where $w(x)$ is given by Eq. (3.2), E denotes Young's modulus and $\gamma'(x)$ is constant due to Eq. (3.4). The plastic slip $\gamma(x) = w'(x)$ may be understood in an averaged sense as visualized in Fig. 3.2.
The total energy is

$$W_\mathrm{tot} = N_\mathrm{c}W_i = \frac{1}{24}L_\mathrm{c}^3 Eh^2(\gamma'(x))^2. \tag{3.10}$$

The average volumetric energy reads

$$\overline{W}_\mathrm{g} = \frac{W_\mathrm{tot}}{L_\mathrm{c}^3} = \frac{1}{2}\frac{E}{12}h^2(\boldsymbol{d}\cdot\nabla\gamma)^2, \tag{3.11}$$

where $\gamma'(x) = \boldsymbol{d}\cdot\nabla\gamma$ has been used.

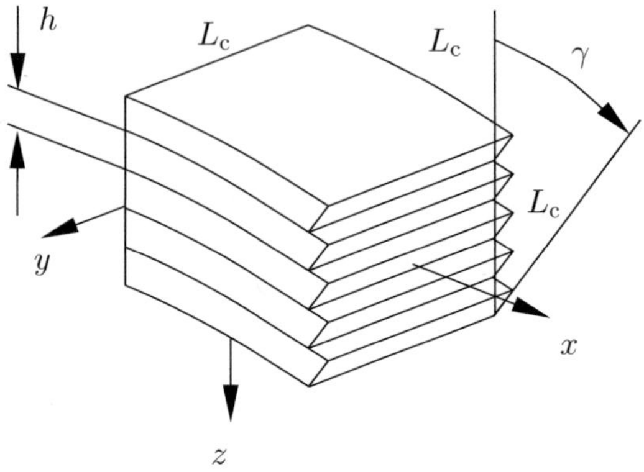

Figure 3.2: Layered cube with strictly localized plastic slip. For illustration reasons, only five layers are shown. In general, the number of layers is assumed to be much larger.

Remarks:

- Note that result (3.11) is independent of L_c, i.e., the result does not change, if the number of layers N_c (and consequently $L_c = N_c h$) is changed as long as h and the curvature $w''(x)$ remain constant. Therefore, the term 'volumetric energy' is indeed meaningful, if a sufficient number of layers is considered.
- Moreover, note that for a fixed layer thickness h the volumetric energy
$$\overline{W}_g \overset{\gamma'=\gamma_0/L_c}{=} \frac{1}{2}\frac{E}{12}\left(\frac{h}{L_c}\right)^2 \gamma_0^2 \qquad (3.12)$$
may be neglected in the case $h \ll L_c$. In particular, in the limit case $h/L_c \to 0$ the model with localized plastic deformation tends to the model with continuous plastic deformation, comp. Eq. (3.8).
- The energy depends quadratically on $\gamma' = \boldsymbol{d} \cdot \nabla\gamma$. This expression represents the bending of the layers. It is also correlated with the geometrically necessary edge dislocation density $\rho_\vdash = -\boldsymbol{d} \cdot \nabla\gamma$ (Eq. (1.24)).

It is concluded that the material with localized slip bands may be modeled by the classical continuum crystal plasticity theory, if the layers

are sufficiently thin. However, in the case of a finite layer thickness h the application of the classical theory implies an error. This error is (approximately) given by the energy $\overline{W}_{\mathrm{g}}$ (Eq. (3.11)).
One might want to extend the classical theory with continuous plastic deformation to materials with localized slip bands, without explicit resolution of the slip bands. Therefore, one might try to modify the stored energy. In the present case, a possible modified stored energy for the continuous model (the model without explicitly resolved, localized slip bands) is given by

$$W(\boldsymbol{\varepsilon}, \gamma, \nabla\gamma) = \frac{1}{2}(\boldsymbol{\varepsilon} - \boldsymbol{\varepsilon}^{\mathrm{p}}) \cdot \mathbb{C}[\boldsymbol{\varepsilon} - \boldsymbol{\varepsilon}^{\mathrm{p}}] + \frac{1}{2}\frac{E}{12}h^2(\boldsymbol{d} \cdot \nabla\gamma)^2. \tag{3.13}$$

If the prescribed displacement field (3.2) and the γ-field (3.4) are applied, the continuous model would then yield the correct energy, which matches the prediction of the model with discontinuous plastic slip. However, it is questionable if the energy (3.13) also holds for more general deformations. This question will be pursued in the subsequent sections. Moreover, the limited accuracy of the applied beam theory will be discussed.

3.2.2 Plastic Torsion

Continuous Plastic Deformation

Now consider an elasto-plastic cube under a torsion-like deformation (Fig. 3.3) with the displacement field

$$\boldsymbol{u}(\boldsymbol{x}) = w(x, y)\boldsymbol{e}_z = \theta_0 \frac{x}{L_{\mathrm{c}}} y \boldsymbol{e}_z \tag{3.14}$$

and displacement gradient

$$\boldsymbol{H} = \frac{\theta_0}{L_{\mathrm{c}}} y \boldsymbol{e}_z \otimes \boldsymbol{e}_x + \theta_0 \frac{x}{L_{\mathrm{c}}} \boldsymbol{e}_z \otimes \boldsymbol{e}_y. \tag{3.15}$$

This deformation can be characterized by the angle θ_0 (Fig. 3.3).
Two slip systems are introduced with common slip plane normal $\boldsymbol{n} = \boldsymbol{e}_z$

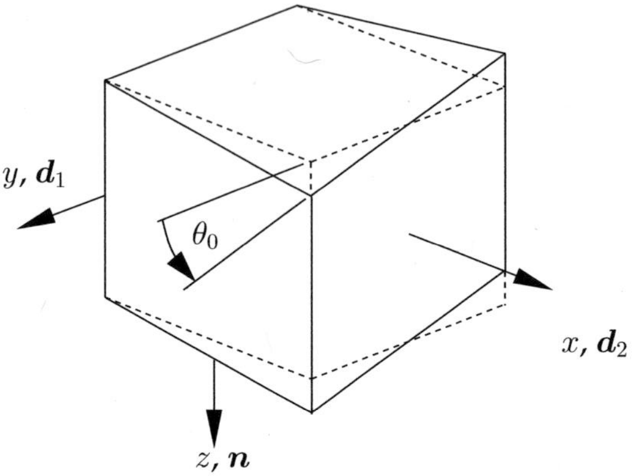

Figure 3.3: Cube suspended to a torsion like deformation.

and orthogonal slip directions $\boldsymbol{d}_1 = \boldsymbol{e}_y$ and $\boldsymbol{d}_2 = \boldsymbol{e}_x$. The plastic slips are prescribed by

$$\gamma_1(x) = \theta_0 \frac{x}{L_\mathrm{c}}, \quad \gamma_2(y) = \frac{\theta_0}{L_\mathrm{c}} y, \tag{3.16}$$

i.e., the plastic part of the displacement gradient reads

$$\boldsymbol{H}^\mathrm{p} = \theta_0 \frac{x}{L_\mathrm{c}} \boldsymbol{e}_y \otimes \boldsymbol{e}_z + \frac{\theta_0}{L_\mathrm{c}} y \boldsymbol{e}_x \otimes \boldsymbol{e}_z. \tag{3.17}$$

Again, the elastic strains and the stress as well as the elastic strain energy vanish

$$\boldsymbol{\varepsilon}^\mathrm{e} = \boldsymbol{\sigma} = \boldsymbol{0} \ \Rightarrow W_\mathrm{e} = 0. \tag{3.18}$$

Localized Plastic Deformation

The localized counterpart of the deformation is realized by thin-walled beams with the same dimensions as in the last section. The deformation is assumed to include warping as illustrated in Fig. 3.4. Otherwise the cross sections perpendicular to the x- and the y-axes are assumed to preserve their rectangular shape (the exact displacement field will be discussed subsequently).
As can be concluded from Fig. 3.4 the average plastic shears are formally

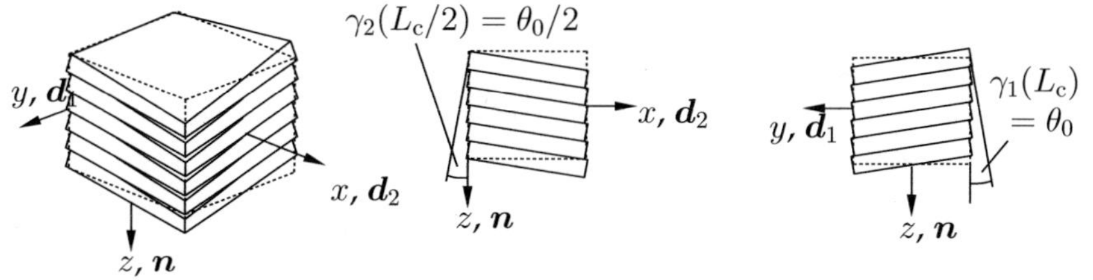

Figure 3.4: Cube suspended to torsion.

identical with the continuous case, Eq. 3.16. The total strain energy of one layer is given by

$$W_i = \frac{1}{2}\int_0^{L_c} GI_T(\theta'(x))^2\,dx = \frac{1}{6}Gh^3\theta_0^2 = \frac{4}{24}Gh^3L_c^2\left(\frac{\theta_0}{L_c}\right)^2 \tag{3.19}$$

with $I_T = 1/3L_ch^3$, $\theta(x) = \theta_0 x/L_c$ and the shear modulus G. The volumetric energy consequently reads in this specific case

$$\overline{W}_g = \frac{W_{tot}}{L_c^3} = \frac{N_cW_i}{L_c^3} = \frac{1}{2}\frac{G}{12}h^2q(\partial_x\gamma_1, \partial_y\gamma_2). \tag{3.20}$$

Here, q is a non-unique function of the plastic strain gradients. A possible realization is given by

$$q = (\boldsymbol{l}_1\cdot\nabla\gamma_1)^2 + (\boldsymbol{l}_2\cdot\nabla\gamma_2)^2 + 2(\boldsymbol{l}_1\cdot\nabla\gamma_1)(\boldsymbol{l}_2\cdot\nabla\gamma_2), \tag{3.21}$$

with $\boldsymbol{l}_1 = \boldsymbol{e}_x$ and $\boldsymbol{l}_2 = \boldsymbol{e}_y$. The form (3.21) will be motivated in the subsequent sections.

Remarks:

- The two slip system directions are orthogonal, which seems to be artificial if, e.g., FCC-crystals are considered. However, the second slip system can be replaced by two systems with directions $\boldsymbol{d}_a = \sqrt{3}/2\boldsymbol{e}_x - 1/2\boldsymbol{e}_y$ and $\boldsymbol{d}_b = -\sqrt{3}/2\boldsymbol{e}_x - 1/2\boldsymbol{e}_y$ with slip plane normal $\boldsymbol{n} = \boldsymbol{e}_z$. If the associated slips are chosen to be $\gamma_a = -\gamma_b = \gamma_2/\sqrt{3}$ the computations yield the same results.
- The projected plastic strain gradients in Eq. (3.21) can be interpreted as geometrically necessary screw dislocation densities (see Eq. (1.24)).

3.2.3 Single-Slip Torsion

The last example investigated the strain gradient influence in the case of two slip systems. It was chosen in order to clearly demonstrate that the continuous theory yields a solution with zero strain energy, while the discontinuous solution does not. Therefore, two active slip systems were required. Consequently, the function $q(\partial_x\gamma_1, \partial_y\gamma_2)$, Eq. (3.21), could not uniquely be chosen. In order to identify a unique energy for one slip system (like in the bending example), the deformation is slightly modified. Therefore, the displacement fields of the layers in the last example (Sect. 3.2.2) are revised more precisely

$$\begin{aligned} \boldsymbol{u}_i(\boldsymbol{x}) &= u_i(y,z)\boldsymbol{e}_x + v_i(x,z)\boldsymbol{e}_y + w_i(x,y)\boldsymbol{e}_z \\ &= -\theta_0\frac{y}{L_\mathrm{c}}\Delta z(z)\boldsymbol{e}_x - \theta_0\frac{x}{L_\mathrm{c}}\Delta z(z)\boldsymbol{e}_y + \theta_0 y\frac{x}{L_\mathrm{c}}\boldsymbol{e}_z. \end{aligned} \tag{3.22}$$

Here, i is the layer index. If z_i represents the z-coordinate of the layer center plane, the function $\Delta z(z) = z - z_i$ describes the (signed) distance from this plane. The strain tensor can be computed by

$$\boldsymbol{\varepsilon}_i = -\frac{2\theta_0}{L_\mathrm{c}}\Delta z(z)\,\mathrm{sym}(\boldsymbol{e}_x \otimes \boldsymbol{e}_y), \tag{3.23}$$

i.e. $\varepsilon_{xy} = \varepsilon_{yx}$ is the only non-vanishing strain component as it is well known from torsion theory. The associated strain energy is given by

$$W_i = \frac{1}{2}\int_{\mathcal{B}_i} \boldsymbol{\sigma}\cdot\boldsymbol{\varepsilon}\,\mathrm{d}v = \frac{1}{6}\theta_0^2 G h^3, \tag{3.24}$$

which coincides with result (3.19), as required. In order to reduce the number of slip systems from two to one, the displacement of the layers in x-direction is constrained to vanish and consistently the plastic slip γ_2 vanishes, too (comp. Fig. 3.4). The index of the first slip system will therefore be dropped in the following. The displacement field reads (comp. Eq. (3.22))

$$\boldsymbol{u}_i(\boldsymbol{x}) = -\theta_0\frac{x}{L_\mathrm{c}}\Delta z(z)\boldsymbol{e}_y + \theta_0 y\frac{x}{L_\mathrm{c}}\boldsymbol{e}_z, \tag{3.25}$$

with strain tensor

$$\varepsilon_i = -\frac{\theta_0}{L_c}\Delta z(z)\,\mathrm{sym}(\boldsymbol{e}_x \otimes \boldsymbol{e}_y) + \frac{\theta_0}{L_c} y\,\mathrm{sym}(\boldsymbol{e}_x \otimes \boldsymbol{e}_z) \tag{3.26}$$

and stain energy

$$W_i = \frac{1}{2}\int_{\mathcal{B}_i} \boldsymbol{\sigma} \cdot \boldsymbol{\varepsilon}\,\mathrm{d}v = \frac{G\theta_0^2}{24}(hL_c^2 + h^3). \tag{3.27}$$

The total strain energy reads

$$W_{\mathrm{tot}} = N_c W_i = L_c^3 \left(\frac{G\theta_0^2}{24} + \frac{1}{2}\frac{G}{12}h^2(\boldsymbol{l} \cdot \nabla\gamma)^2\right). \tag{3.28}$$

Comparing this result with the result from the continuous model based on the displacement field (3.14) and $\boldsymbol{H}^{\mathrm{p}} = \gamma(x)\boldsymbol{e}_y \otimes \boldsymbol{e}_z$ (with $\gamma(x) = \theta_0 x / L_c$, Eq. $(3.16)_1$)

$$W_{\mathrm{tot}} = L_c^3 \frac{G\theta_0^2}{24} \tag{3.29}$$

clearly illustrates the influence of a finite layer thickness incorporated by the strain gradient term in Eq. (3.28). Again, consistency is reached for $h/L_c \to 0$. The volumetric energy associated to the plastic strain gradient influence for this example is given by

$$\overline{W}_{\mathrm{g}} = \frac{1}{2}\frac{G}{12}h^2(\boldsymbol{l} \cdot \nabla\gamma)^2, \tag{3.30}$$

with $\boldsymbol{l} = -\boldsymbol{e}_x$. Alternatively, the energy (3.30) can be written as

$$\overline{W}_{\mathrm{g}} = \frac{1}{2}\frac{G}{12}h^2(\boldsymbol{h} \cdot \nabla\gamma)^2, \tag{3.31}$$

with $\boldsymbol{h} = \boldsymbol{e}_x$ (see Fig. 1.2). At this point the choice of the function $q(\partial_x\gamma_1, \partial_y\gamma_2)$ (Eq. (3.21)) becomes clear, since it is consistent with Eq. (3.30).

3.2.4 Need for a new Continuum Theory

The bending and torsion examples illustrate that materials with localized slip bands might be modeled by a classical continuum crystal theory. This approach seems reasonable if the slip band distances are negligible compared to the wavelengths of the overall deformation. However, in the case of finite layer thicknesses the energy of the model with discrete slip bands is not reasonably predicted by the continuous model. In order to extend the classical crystal plasticity theory, energy correction terms might be introduced. The bending and torsion examples suggest the following approach for the single slip case

$$W = \frac{1}{2}\boldsymbol{\varepsilon}^{\mathrm{e}} \cdot \mathbb{C}[\boldsymbol{\varepsilon}^{\mathrm{e}}] + \frac{1}{2}\frac{h^2}{12}\left(E(\boldsymbol{d} \cdot \nabla\gamma)^2 + G(\boldsymbol{l} \cdot \nabla\gamma)^2\right). \qquad (3.32)$$

This energy approach implies energy predictions which are consistent with the discontinuous bending and torsion examples. However, the generality of approach (3.32) is questionable. Therefore, more general deformation modes will be considered in the subsequent section.

3.3 Generalized Energy Functions

3.3.1 Kinematical Assumptions

Microscopic Fields

Subsequently, an estimation of the stored energy density is derived for more general deformation types than bending and torsion. More precisely, several hierarchical energies are developed for the single slip case. First, a kinematical framework based on a microscopic displacement field $\boldsymbol{u}^*(\boldsymbol{x})$ is formulated in order to evaluate the microscopic elastic energy. This energy is taken as a function of the microscopic strains.

Assumptions on the microscopic displacement field $\boldsymbol{u}^*(\boldsymbol{x})$:

- The displacement field $\boldsymbol{u}^*(\boldsymbol{x})$ is defined on the micro-scale taking the slip bands into account explicitly as ideal displacement jumps $[\![\boldsymbol{u}^*(\boldsymbol{x})]\!] = \boldsymbol{u}^+ - \boldsymbol{u}^-$, with the normal $\boldsymbol{n}$ pointing from '-' to '+' (see Fig. 3.5).

- The direction of the displacement jumps is given by the slip direction $\boldsymbol{d}$, i.e.,

$$\boldsymbol{l} \cdot [\![\boldsymbol{u}^*(\boldsymbol{x})]\!] = \boldsymbol{n} \cdot [\![\boldsymbol{u}^*(\boldsymbol{x})]\!] = 0. \tag{3.33}$$

- It should be noted that the microscopic displacement field $\boldsymbol{u}^*(\boldsymbol{x})$ is *not* defined on some representative volume element. Instead, it is defined throughout the *whole* body $\mathcal{B}$, i.e.

$$\boldsymbol{u}^* = \boldsymbol{u}^*(\boldsymbol{x}), \quad \boldsymbol{x} \in \mathcal{B}. \tag{3.34}$$

- The slip bands are assumed plane and parallel (see Fig. 3.5). They are denoted by $\mathcal{SB}_i$ ($i = 1, ..., N_{\mathcal{SB}}$). Their normal is denoted by $\boldsymbol{n}$. The totality of all slip bands is given by

$$\mathcal{SB} = \bigcup_{i=1}^{N_{\mathcal{SB}}} \mathcal{SB}_i. \tag{3.35}$$

- The distance h between the slip bands is assumed to be constant (Fig. 3.5).
- The regions between the slip bands are referred to as layers with associated center planes $\mathcal{L}_i$ ($i = 1, ... N_{\mathcal{L}}$) (Fig. 3.5). The plastic deformation in the layers is assumed negligible. As a consequence, the elastic strain in the layers is equal to the total strain.
- The localized plastic deformation is assumed to be caused by dislocations moving in the localized slip bands (the dislocations are not explicitly resolved). Moreover, the in-plane dislocation density is assumed large, i.e., the average dislocation spacing in the slip bands is assumed much smaller than the slip band distance h.

Clearly, the aforementioned assumptions constitute a strong idealization of the actual crystal deformation observed in experiments. Although the model is intended to illustrate different aspects of gradient plasticity rather than representing any real material, a discussion of its similarities and differences compared to the experimentally observed single crystal

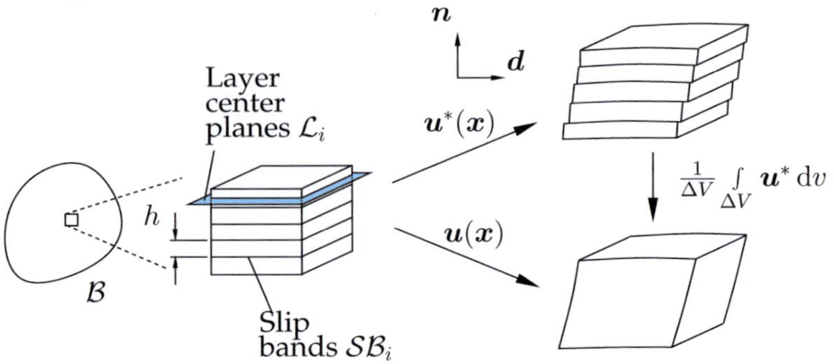

Figure 3.5: Schematic illustration of the mesoscopic and microscopic displacement fields for single slip. Note that the volume ΔV should not be confused with the illustrated cube.

behavior should not be left out.
Experimental investigations of copper single crystals suspended to tension were carried out for example by Mader (1957). Surface observations show that stage I of the deformation is characterized by homogeneously distributed slip lines (fine slip). These slip lines represent strongly localized plastic shear deformations of roughly 10 Burgers vectors length. They are assumed to be the result of a sequence of microscopic strain events. This means that the microscopic plastic deformation is not only discrete in space but also in time. The slip line distribution is nearly homogeneous during stage I.
In stage II, the heterogeneity of the slip line distribution increases. The slip lines tend to group, i.e., they form shear bands (structurized fine slip). Again, the activity of the shear bands seems to be limited in time (Mader, 1957). In stage III, the heterogeneity is again increased. Distinct shear bands become visible.
The modeling approach at hand completely neglects that plasticity is a time discrete phenomenon on the microscale. The localization into slip lines and shear bands is strongly idealized.
In the following, a microscopic kinematical framework based on the aforementioned assumptions is developed. To this end, all microscopic

quantities are denoted as $(\bullet)^*$. For example, the microscopic displacement gradient in the layers is given by $\boldsymbol{H}^*(\boldsymbol{x}) = \nabla \boldsymbol{u}^*(\boldsymbol{x})$ and the microscopic strain tensor reads $\boldsymbol{\varepsilon}^*(\boldsymbol{x}) = \mathrm{sym}(\boldsymbol{H}^*(\boldsymbol{x}))$. The aforementioned displacement jumps $[\![\boldsymbol{u}^*(\boldsymbol{x})]\!]$ represent singularities. They take finite values in the slip bands $\mathcal{SB}_i$ and vanish in the layers.

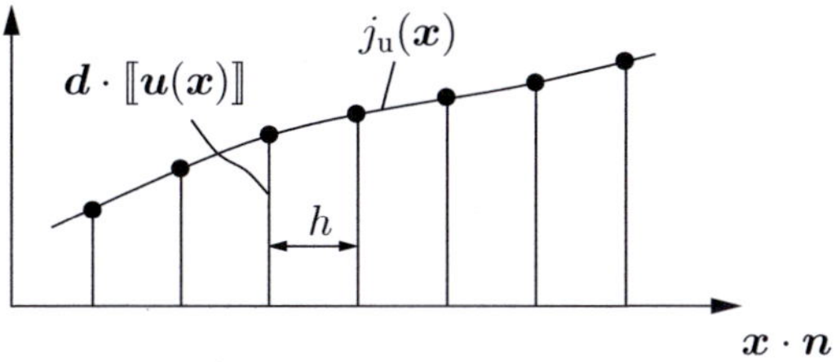

Figure 3.6: Illustration of the interpolation $j_{\mathrm{u}}(\boldsymbol{x})$ of the displacement jumps (along a line perpendicular to the slip bands).

A smooth interpolation field $j_{\mathrm{u}}(\boldsymbol{x})$ is introduced by setting

$$j_{\mathrm{u}}(\boldsymbol{x}) = \begin{cases} \boldsymbol{d} \cdot [\![\boldsymbol{u}^*(\boldsymbol{x})]\!], & \boldsymbol{x} \in \mathcal{SB} \\ \mathrm{itp}\big([\![\boldsymbol{u}^*(\boldsymbol{x})]\!]\big), & \text{otherwise} \end{cases} \tag{3.36}$$

Here, $\mathrm{itp}\big([\![\boldsymbol{u}^*(\boldsymbol{x})]\!]\big)$ is a smooth interpolation of the displacement jumps which guarantees that $j_{\mathrm{u}}(\boldsymbol{x})$ is a continuous and differentiable function everywhere. Fig. 3.6 illustrates the function $j_{\mathrm{u}}(\boldsymbol{x})$. The definition of $j_{\mathrm{u}}(\boldsymbol{x})$ is not unique which leaves some freedom in its choice.

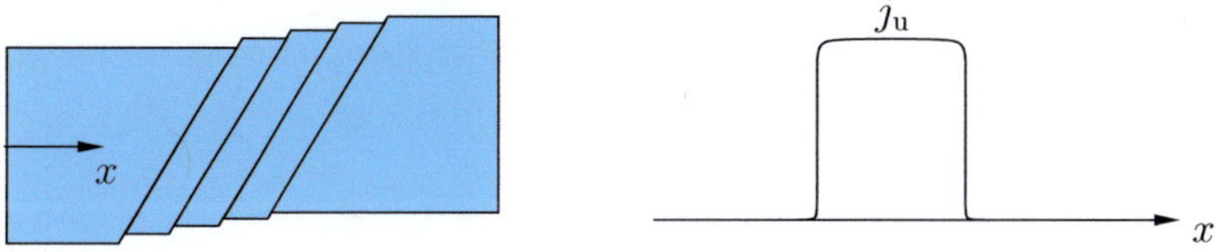

Figure 3.7: Schematic illustration of the function j_{u} along the center line of a bar.

It should be noted that the function $j_{\mathrm{u}}(\boldsymbol{x})$ can in general not be expected to be as smooth as depicted in Fig. 3.6. In particular, variations of $j_{\mathrm{u}}(\boldsymbol{x})$ may be strong in the direction of the slip plane normal $\boldsymbol{n}$. For example, Fig. 3.7 illustrates a case where this situation arises.

Additionally to $j_u(\boldsymbol{x})$, the (signed) distance $\Delta z(\boldsymbol{x})$ of a point $\boldsymbol{x}$ from the closest layer center plane $\mathcal{L}_i$ is defined (cf. Fig. 3.8). This allows for the formulation of an orthogonal projection of a given point $\boldsymbol{x}$ onto the closest layer center plane $\mathcal{L}_i$

$$\boldsymbol{x}^{\mathrm{P}}(\boldsymbol{x}) = \boldsymbol{x} - \Delta z(\boldsymbol{x})\boldsymbol{n}. \tag{3.37}$$

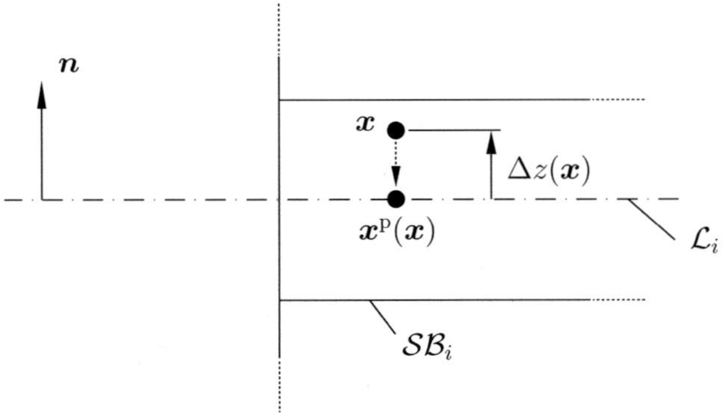

Figure 3.8: Signed distance $\Delta z(\boldsymbol{x})$ in direction $\boldsymbol{n}$ from the closest layer center plane $\mathcal{L}_i$.

Mesoscopic Fields

The microscopic displacement field $\boldsymbol{u}^*(\boldsymbol{x})$ explicitly takes the slip bands into account as displacement jumps. These jumps are eliminated by a smoothing operation in order to obtain the mesoscopic displacement field

$$\boldsymbol{u}(\boldsymbol{x}) = \frac{1}{\Delta V(\boldsymbol{x})} \int_{\Delta V(\boldsymbol{x})} \boldsymbol{u}^*(\boldsymbol{x}')\,\mathrm{d}v', \tag{3.38}$$

where $\Delta V(\boldsymbol{x})$ is an averaging volume with centroid at $\boldsymbol{x}$ and $\mathrm{d}v'$ is the infinitesimal volume element at the position $\boldsymbol{x}'$ (see Fig. 3.9).
It turns out that an advantageous definition of $\Delta V(\boldsymbol{x})$ is not necessarily given by a volume which has similar dimensions in all three spatial directions (like the volume in Fig. 3.9). Instead, the dimension normal to

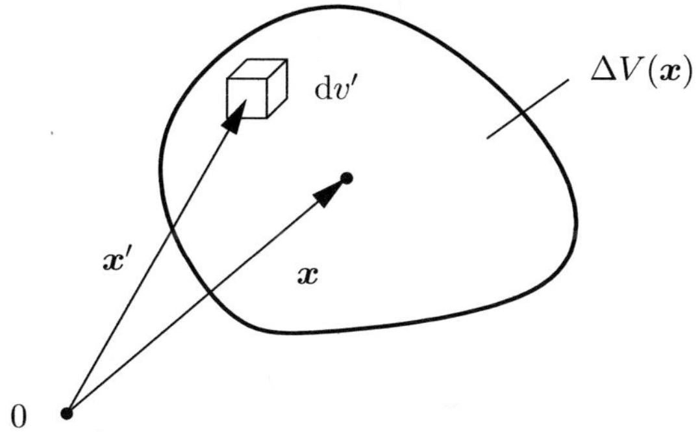

Figure 3.9: Averaging volume with centroid at $\boldsymbol{x}$.

the slip plane is here chosen to coincide with the slip band spacing h while the in-plane dimensions are chosen to be much larger than h (see Fig. 3.10). Let L denote a characteristic in-plane dimension of the averaging volume $\Delta V(\boldsymbol{x})$. The main motivation of this choice of $\Delta V(\boldsymbol{x})$ is the desire to capture strong plastic slip gradients in the out-of-plane direction (as illustrated, e.g., in Fig. 3.7) on the mesoscale.

Fig. 3.5 illustrates the deformation on the micro- and the mesoscale. Moreover, the mesoscopic displacement gradient and strain tensor read $\boldsymbol{H}(\boldsymbol{x}) = \nabla \boldsymbol{u}(\boldsymbol{x})$ and $\boldsymbol{\varepsilon}(\boldsymbol{x}) = \mathrm{sym}(\boldsymbol{H}(\boldsymbol{x}))$, respectively.

The difference of the microscopic and the mesoscopic displacements is given by a micro fluctuation $\tilde{\boldsymbol{u}}(\boldsymbol{x})$

$$\boldsymbol{u}^*(\boldsymbol{x}) = \boldsymbol{u}(\boldsymbol{x}) + \tilde{\boldsymbol{u}}(\boldsymbol{x}). \tag{3.39}$$

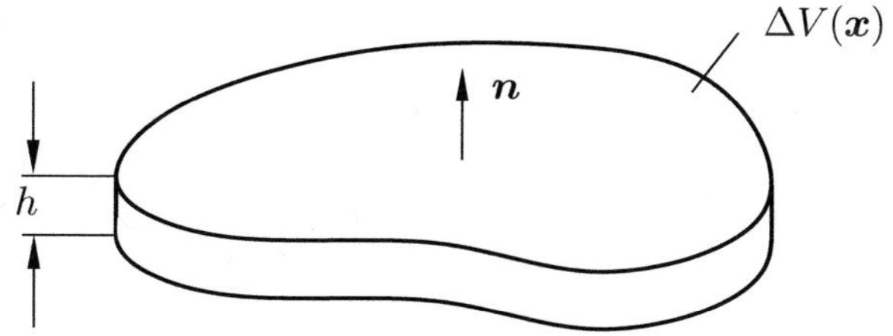

Figure 3.10: Averaging volume $\Delta V(\boldsymbol{x})$.

In addition, the gradient of the displacement fluctuations reads $\tilde{\boldsymbol{H}}(\boldsymbol{x}) = \nabla\tilde{\boldsymbol{u}}(\boldsymbol{x})$ (the properties of $\tilde{\boldsymbol{H}}$ are discussed subsequently). The following expression is adopted as definition of the plastic shear strain

$$\gamma(\boldsymbol{x}) = \frac{1}{\Delta V(\boldsymbol{x})} \int_{\Delta V(\boldsymbol{x})} \delta_{\mathcal{SB}}(\boldsymbol{x}')\, \boldsymbol{d} \cdot [\![\boldsymbol{u}^*(\boldsymbol{x}')]\!]\, \mathrm{d}v'. \tag{3.40}$$

Here, the delta distribution $\delta_{\mathcal{SB}}$ is implicitly defined (for an arbitrary function $f(\boldsymbol{x})$) by

$$\int_{\Delta V(\boldsymbol{x})} \delta_{\mathcal{SB}}(\boldsymbol{x}')\, f(\boldsymbol{x}')\, \mathrm{d}v' = \int_{A} f(\boldsymbol{x}')\, \mathrm{d}a', \tag{3.41}$$

where the short notation $A = \Delta V(\boldsymbol{x}) \cap \mathcal{SB}$ has been introduced, for convenience.

3.3.2 A Hierarchy of Energies

The scope of this section is the derivation of several mesoscopic energy density functions with with different levels of accuracy. The functions are supposed to capture the elastic strain energy density of the elasto-plastic material with highly localized plastic deformations in the form of localized slip bands, as introduced in the previous section. The energies are defined on the mesoscopic scale in terms of the mesoscopic displacement field $\boldsymbol{u}(\boldsymbol{x})$ and the plastic shear strain $\gamma(\boldsymbol{x})$ (and their gradients). The average energy density of a given averaging volume $\Delta V(\boldsymbol{x})$ is given by

$$W(\boldsymbol{x}) = \frac{1}{\Delta V(\boldsymbol{x})} \int_{\Delta V(\boldsymbol{x})} \frac{1}{2} \boldsymbol{\varepsilon}^*(\boldsymbol{x}') \cdot \mathbb{C}[\boldsymbol{\varepsilon}^*(\boldsymbol{x}')]\, \mathrm{d}v'. \tag{3.42}$$

Here and in the following, the centroid $\boldsymbol{x}$ of $\Delta V(\boldsymbol{x})$ is treated as a fixed but otherwise arbitrary point. Due to the minor symmetry of the stiffness tensor $\mathbb{C}$, the strain can be replaced by the displacement gradient

$$\boldsymbol{\varepsilon}^*(\boldsymbol{x}') \cdot \mathbb{C}[\boldsymbol{\varepsilon}^*(\boldsymbol{x}')] = \boldsymbol{H}^*(\boldsymbol{x}') \cdot \mathbb{C}[\boldsymbol{H}^*(\boldsymbol{x}')]. \tag{3.43}$$

The microscopic displacement gradient can be expressed in terms of the mesoscopic gradient and the fluctuation gradient

$$\boldsymbol{H}^*(\boldsymbol{x}') = \boldsymbol{H}(\boldsymbol{x}') + \tilde{\boldsymbol{H}}(\boldsymbol{x}'). \tag{3.44}$$

Subsequently, different approaches of $\boldsymbol{H}(\boldsymbol{x}')$ and $\tilde{\boldsymbol{H}}(\boldsymbol{x}')$ are introduced in order to derive approximations of $W(\boldsymbol{x})$ (Eq.(3.42)).

A Simple Energy Based on Constant Strain Approximations

In order to derive a simple approximation W_{11} (the meaning of the indices will subsequently be explained) of W, the Taylor series expansion of the mesoscopic displacement gradient is truncated after the constant term

$$\boldsymbol{H}(\boldsymbol{x}') \approx \boldsymbol{H}_1(\boldsymbol{x}') = \boldsymbol{H}(\boldsymbol{x}), \tag{3.45}$$

where $\boldsymbol{x}$ denotes the centroid of the averaging volume $\Delta V(\boldsymbol{x})$ (Eq. (3.42)). The index '1' indicates that the Taylor series expansion of $\boldsymbol{H}$ is truncated after the first (constant) term. For the displacement field it follows that

$$\boldsymbol{u}(\boldsymbol{x}') \approx \boldsymbol{u}_1(\boldsymbol{x}') = \boldsymbol{u}(\boldsymbol{x}) + \boldsymbol{H}(\boldsymbol{x})(\boldsymbol{x}' - \boldsymbol{x}). \tag{3.46}$$

In order to derive an approximation of the fluctuations, the following ansatz is proposed

$$\tilde{\boldsymbol{u}}(\boldsymbol{x}') = -j_{\mathrm{u}}\big(\boldsymbol{x}^{\mathrm{P}}(\boldsymbol{x}')\big)\, \boldsymbol{d}\, \frac{\Delta z(\boldsymbol{x}')}{h}. \tag{3.47}$$

This approach implies that the fluctuations in the layers are linear with respect to Δz, as illustrated in Fig. 3.11.
For the computation of $W_{11}(\boldsymbol{x})$, the function $j_{\mathrm{u}}(\boldsymbol{x}')$ is assumed to be constant in $\Delta V(\boldsymbol{x})$ (Eq. (3.42)). This represents an analogue of a Taylor series expansion truncated after the first term

$$j_{\mathrm{u}}(\boldsymbol{x}') \approx j_{\mathrm{u},1}(\boldsymbol{x}') = j_{\mathrm{u}0} \Rightarrow \quad [\![\boldsymbol{u}^*(\boldsymbol{x}')]\!] \approx j_{\mathrm{u}0}\boldsymbol{d}, \;\; \boldsymbol{x}' \in \mathcal{SB} \cap \Delta V(\boldsymbol{x}). \tag{3.48}$$

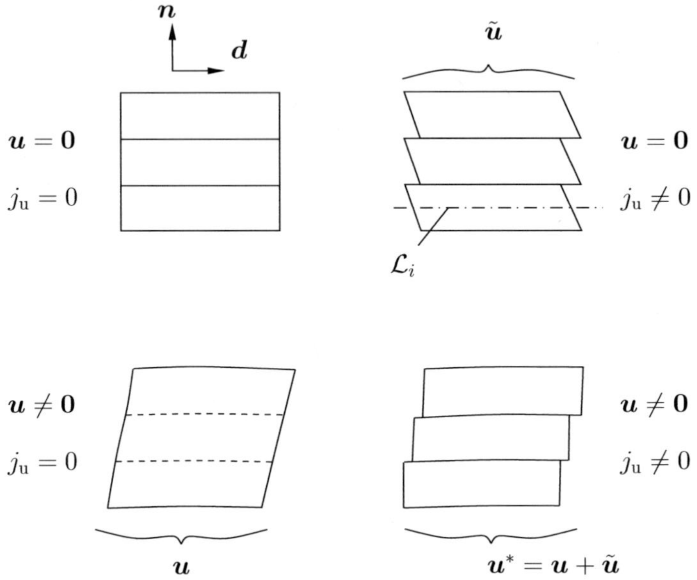

Figure 3.11: The micro displacement field $\boldsymbol{u}^*(\boldsymbol{x})$ is a superposition of the mesoscopic field $\boldsymbol{u}(\boldsymbol{x})$ and the fluctuations $\tilde{\boldsymbol{u}}(\boldsymbol{x})$.

The meaning of the index of $j_{u,1}(\boldsymbol{x}')$ is analogous to the index of $\boldsymbol{H}_1(\boldsymbol{x}')$ (truncation after the constant term). The unknown constant j_{u0} can be expressed in terms of $\gamma(\boldsymbol{x})$ owing to the following consistency requirement (cf. the definition of $\gamma(\boldsymbol{x})$, Eq. (3.40))

$$\gamma(\boldsymbol{x}) \stackrel{!}{=} \frac{1}{\Delta V(\boldsymbol{x})} \int_{\Delta V(\boldsymbol{x})} j_{u0} \delta_{\mathcal{SB}}(\boldsymbol{x}') \, dv' = \frac{j_{u0}}{h}. \tag{3.49}$$

Eq. (3.49) leads to the following representation of the fluctuations (Eq. (3.47))

$$\tilde{\boldsymbol{u}}_1(\boldsymbol{x}') = -\Delta z(\boldsymbol{x}')\gamma(\boldsymbol{x})\boldsymbol{d}. \tag{3.50}$$

As a result, the microscopic displacement gradient is approximated by

$$\boldsymbol{H}^*_{11}(\boldsymbol{x}') = \boldsymbol{H}_1(\boldsymbol{x}') + \tilde{\boldsymbol{H}}_1(\boldsymbol{x}') \stackrel{(3.45),(3.50)}{=} \boldsymbol{H}(\boldsymbol{x}) - \gamma(\boldsymbol{x})\boldsymbol{d} \otimes \boldsymbol{n}. \tag{3.51}$$

Here, the first index of $\boldsymbol{H}^*_{11}$ represents the index of $\boldsymbol{H}_1(\boldsymbol{x}')$, the second index is associated with the index of $\tilde{\boldsymbol{H}}_1(\boldsymbol{x}')$. The expression on the right hand side of Eq. (3.51) is well known from continuum crystal plasticity theories as the elastic part of the displacement gradient

$$\boldsymbol{H}^*_{11}(\boldsymbol{x}') = \boldsymbol{H}^{\mathrm{e}}(\boldsymbol{x}) = \boldsymbol{H}(\boldsymbol{x}) - \gamma(\boldsymbol{x})\boldsymbol{d} \otimes \boldsymbol{n}. \tag{3.52}$$

Finally, the energy $W(\boldsymbol{x})$ (Eq. (3.42)) can be approximated by insertion of the identity $\boldsymbol{H}^*_{11}(\boldsymbol{x}') = \boldsymbol{H}^{\mathrm{e}}(\boldsymbol{x})$ into Eq. (3.42) and letting $\Delta V(\boldsymbol{x}) \to 0$[6]

$$W_{11}(\boldsymbol{x}) = \lim_{\Delta V(\boldsymbol{x}) \to 0} \frac{1}{\Delta V(\boldsymbol{x})} \int_{\Delta V(\boldsymbol{x})} \frac{1}{2} \boldsymbol{H}^*_{11}(\boldsymbol{x}) \cdot \mathbb{C}[\boldsymbol{H}^*_{11}(\boldsymbol{x})] \, \mathrm{d}v'. \tag{3.53}$$

The indices of W_{11} represent the accuracy of the approach. They match the indices of $\boldsymbol{H}^*_{11}$. Summarizing the results, the simplest possible approximations of $\boldsymbol{H}(\boldsymbol{x}')$ and of the fluctuations (in terms of $j_{\mathrm{u}}(\boldsymbol{x})$) leads to the classical energy

$$W_{11}(\boldsymbol{x}) = \frac{1}{2} \boldsymbol{\varepsilon}^{\mathrm{e}}(\boldsymbol{x}) \cdot \mathbb{C}[\boldsymbol{\varepsilon}^{\mathrm{e}}(\boldsymbol{x})], \tag{3.54}$$

with $\boldsymbol{\varepsilon}^{\mathrm{e}}(\boldsymbol{x}) = \mathrm{sym}(\boldsymbol{H}^{\mathrm{e}}(\boldsymbol{x}))$.

Clearly, the energy density W_{11} can be applied in situations where the characteristic wavelengths of the deformation field are much larger than the layer thickness h. In this situation, the microscopic strain is nearly homogeneous in the averaging volume $\Delta V(\boldsymbol{x})$.

It remains to be shown, that the fluctuation approach (3.50) is consistent with the definition of $\boldsymbol{u}(\boldsymbol{x})$ (Eq. (3.38)). This is done in App. B.1.

Enhanced Displacement Approximation

As has been discussed in the previous section, the energy density W_{11} represents a reasonable mesoscopic approach, if the characteristic wavelengths of the mesoscopic deformation field are much larger than the layer thickness h. In order to derive a more accurate energy than $W_{11}(\boldsymbol{x})$, the approximation of the mesoscopic displacement gradient is improved. As mentioned in Sect. 3.3.1, the characteristic in-plane

[6]To the authors' knowledge, there is no theorem which guarantees the reasonability of the energy densities which are derived from an approach like (3.53). Therefore, each subsequently computed energy density must and will be verified and checked concerning its plausibility.

dimension L of the averaging volume is assumed significantly larger than the normal dimension h. Therefore, the mesoscopic displacement gradient $\boldsymbol{H}$ is approximated linearly in the slip plane and constant normal to the slip plane, for simplicity, by taking into account the linear in-plane term of the Taylor series expansion

$$\boldsymbol{H}_2(\boldsymbol{x}') = \boldsymbol{H}(\boldsymbol{x}) + \nabla_{\mathrm{p}}\boldsymbol{H}(\boldsymbol{x})(\boldsymbol{x}' - \boldsymbol{x}) = \boldsymbol{H}(\boldsymbol{x}) + \nabla\boldsymbol{H}(\boldsymbol{x})(\boldsymbol{I} - \boldsymbol{n}\otimes\boldsymbol{n})(\boldsymbol{x}' - \boldsymbol{x}). \tag{3.55}$$

Here, the gradient is given by $\nabla\boldsymbol{H} = \partial_{x_k} H_{ij}\boldsymbol{e}_i \otimes \boldsymbol{e}_j \otimes \boldsymbol{e}_k$.
The constant approximation (3.48) of the fluctuations is applied again. For that reason, Eq. (3.50) remains valid

$$\tilde{\boldsymbol{u}}_1(\boldsymbol{x}') = -\Delta z(\boldsymbol{x}')\gamma(\boldsymbol{x})\boldsymbol{d}. \tag{3.56}$$

As a consequence, the microscopic displacement gradient reads (cf. Eq. (3.44))

$$\begin{aligned} \boldsymbol{H}^*_{21}(\boldsymbol{x}') &= \boldsymbol{H}_2(\boldsymbol{x}') + \tilde{\boldsymbol{H}}_1(\boldsymbol{x}') && (3.57)\\ &= \boldsymbol{H}(\boldsymbol{x}) - \gamma(\boldsymbol{x})\boldsymbol{d}\otimes\boldsymbol{n} + \nabla_{\mathrm{p}}\boldsymbol{H}(\boldsymbol{x})(\boldsymbol{x}' - \boldsymbol{x}). && (3.58) \end{aligned}$$

The associated energy $W_{21}(\boldsymbol{x})$ is defined by

$$W_{21}(\boldsymbol{x}) = \lim_{\Delta V(\boldsymbol{x})\to 0} \frac{1}{\Delta V(\boldsymbol{x})} \int_{\Delta V(\boldsymbol{x})} \frac{1}{2}\boldsymbol{H}^*_{21}(\boldsymbol{x}') \cdot \mathbb{C}[\boldsymbol{H}^*_{21}(\boldsymbol{x}')]\, \mathrm{d}v'. \tag{3.59}$$

The combination of Eqns. (3.57) and (3.59) and yields

$$\begin{aligned} W_{21}(\boldsymbol{x}) &= \lim_{\Delta V(\boldsymbol{x})\to 0} \left(\frac{1}{2}\boldsymbol{H}^{\mathrm{e}}(\boldsymbol{x}) \cdot \mathbb{C}[\boldsymbol{H}^{\mathrm{e}}(\boldsymbol{x})] + \mathcal{O}(L^2)\right) && (3.60)\\ &= \frac{1}{2}\boldsymbol{\varepsilon}^{\mathrm{e}}(\boldsymbol{x}) \cdot \mathbb{C}[\boldsymbol{\varepsilon}^{\mathrm{e}}(\boldsymbol{x})], && (3.61) \end{aligned}$$

with $\boldsymbol{H}^{\mathrm{e}}$ as defined in Eq. (3.52). Obviously, the improvement of the previous ansatz $\boldsymbol{H}_{11}(\boldsymbol{x})$ (Eq. (3.51)) based on the consideration of the mesoscopic in-plane gradient $\nabla_{\mathrm{p}}\boldsymbol{H}(\boldsymbol{x})$ does not alter the result, i.e.

$$W_{21}(\boldsymbol{x}) = W_{11}(\boldsymbol{x}). \tag{3.62}$$

This result is verified in App. B.2.

Remarks:

- In the purely elastic case (i.e. $\gamma(\boldsymbol{x}) = 0$), the microscopic and mesoscopic constitutive equations of the material should coincide, since there is no internal length scale separating the two scales. Hence, the only consistent way to include $\nabla_{\rm p}\boldsymbol{H}$ into the energy consists of a $\nabla_{\rm p}\boldsymbol{H}$-dependent term, which vanishes in the elastic case. Therefore, the only meaningful mesoscopic energy approach for the purely elastic case is given by

$$\frac{1}{2}\boldsymbol{\varepsilon}(\boldsymbol{x}) \cdot \mathbb{C}[\boldsymbol{\varepsilon}(\boldsymbol{x})]. \tag{3.63}$$

  This condition is fulfilled by the approximation $W_{21}(\boldsymbol{x})$ (Eq. (3.61)).

An Energy Including Plastic Strain Gradients

In order to derive an improved energy, the constant ansatz (3.48) for $j_{\rm u}$ is replaced by a linear approach. Again, this ansatz represents an analogue of a Taylor series expansion, now truncated after the linear term

$$j_{\rm u,2}(\boldsymbol{x}') \approx j_{\rm u0} + \boldsymbol{b}_0 \cdot (\boldsymbol{x}' - \boldsymbol{x}). \tag{3.64}$$

Since the out-of-plane dimension h of the averaging volume $\Delta V(\boldsymbol{x})$ is significantly smaller than the characteristic in-plane dimension L, the out-of-plane component of $\boldsymbol{b}_0$ is neglected for simplicity

$$\boldsymbol{n} \cdot \boldsymbol{b}_0 = 0. \tag{3.65}$$

Again, the fluctuations are approximated by ansatz (3.47), given by

$$\tilde{\boldsymbol{u}}(\boldsymbol{x}') = -j_{\rm u}\big(\boldsymbol{x}^{\rm p}(\boldsymbol{x}')\big)\,\boldsymbol{d}\,\frac{\Delta z(\boldsymbol{x}')}{h}. \tag{3.66}$$

In App. B.3, it is shown that this ansatz (in combination with Eq. (3.64)) is consistent with the definition (3.36) of $j_{\rm u}(\boldsymbol{x})$.
The constants $j_{\rm u0}$ and $\boldsymbol{b}_0$ are determined based on consistency conditions

with respect to the field $\gamma(\boldsymbol{x})$. First, an approximation of $\gamma(\boldsymbol{x}')$ is computed by inserting ansatz (3.64) in the definition of γ (Eq. (3.40))

$$\gamma_{\text{appr.}}(\boldsymbol{x}') = \frac{1}{\Delta V(\boldsymbol{x}')} \int_{\Delta V(\boldsymbol{x}')} \big(j_{\text{u}0} + \boldsymbol{b}_0 \cdot (\boldsymbol{x}'' - \boldsymbol{x})\big)\delta_{\mathcal{SB}}\,\mathrm{d}v'' \tag{3.67}$$

$$= \frac{1}{h}\big(j_{\text{u}0} + \boldsymbol{b}_0 \cdot (\boldsymbol{x}' - \boldsymbol{x})\big). \tag{3.68}$$

The first consistency condition reads

$$\gamma(\boldsymbol{x}) \overset{!}{=} \gamma_{\text{appr}}(\boldsymbol{x}). \tag{3.69}$$

Consequently, the constant $j_{\text{u}0}$ is given by

$$j_{\text{u}0} = h\,\gamma(\boldsymbol{x}). \tag{3.70}$$

The second condition is given by

$$(\boldsymbol{I} - \boldsymbol{n} \otimes \boldsymbol{n})\nabla\gamma(\boldsymbol{x}) = (\boldsymbol{I} - \boldsymbol{n} \otimes \boldsymbol{n})\partial_{\boldsymbol{x}}\gamma(\boldsymbol{x}) \overset{!}{=} (\boldsymbol{I} - \boldsymbol{n} \otimes \boldsymbol{n})\partial_{\boldsymbol{x}'}\gamma_{\text{appr}}(\boldsymbol{x}) = \frac{\boldsymbol{b}_0}{h}. \tag{3.71}$$

Here, the normal component of the gradient has been neglected as a result of assumption (3.65). Eq. (3.71) yields

$$\boldsymbol{b}_0 = h\,(\boldsymbol{I} - \boldsymbol{n} \otimes \boldsymbol{n})\nabla\gamma(\boldsymbol{x}) = h\nabla_{\text{p}}\gamma(\boldsymbol{x}). \tag{3.72}$$

Here, $\nabla_{\text{p}}\gamma(\boldsymbol{x}) = (\boldsymbol{I} - \boldsymbol{n} \otimes \boldsymbol{n})\nabla\gamma(\boldsymbol{x})$ represents the gradient of $\gamma(\boldsymbol{x})$ projected onto the glide plane. This quantity can be reformulated

$$\nabla_{\text{p}}\gamma(\boldsymbol{x}) = (\boldsymbol{d} \cdot \nabla\gamma(\boldsymbol{x}))\boldsymbol{d} \; + \; (\boldsymbol{h} \cdot \nabla\gamma(\boldsymbol{x}))\boldsymbol{h}, \tag{3.73}$$

with $\boldsymbol{h} = \boldsymbol{d} \times \boldsymbol{n}$ (see Fig. 1.2). Eqns. (3.70) and (3.72) allow to reformulate the fluctuations (Eq. (3.66)) in terms of $\gamma(\boldsymbol{x})$

$$\tilde{\boldsymbol{u}}_2(\boldsymbol{x}') = -\Delta z(\boldsymbol{x}')\gamma_2(\boldsymbol{x}^{\text{P}}(\boldsymbol{x}'))\,\boldsymbol{d}, \tag{3.74}$$

with

$$\gamma_2(\boldsymbol{x}^{\text{P}}(\boldsymbol{x}')) = \gamma(\boldsymbol{x}) + (\boldsymbol{x}' - \Delta z(\boldsymbol{x}')\boldsymbol{n} - \boldsymbol{x}) \cdot \nabla_{\text{p}}\gamma(\boldsymbol{x}) = \gamma(\boldsymbol{x}) + (\boldsymbol{x}' - \boldsymbol{x}) \cdot \nabla_{\text{p}}\gamma(\boldsymbol{x}). \tag{3.75}$$

Here, Eq. (3.64) has been exploited. Approach (3.74) implies

$$\frac{1}{\Delta V(\boldsymbol{x}')} \int\limits_{\Delta V(\boldsymbol{x}')} \tilde{\boldsymbol{u}}_2(\boldsymbol{x}'')\,\mathrm{d}v'' = 0, \tag{3.76}$$

as required. Having obtained the representation (3.74) of the fluctuation, an approximation of the mesoscopic displacement $\boldsymbol{u}(\boldsymbol{x}')$ is also required to compute the energy density approximation. The Taylor series expansion of $\boldsymbol{H}(\boldsymbol{x}')$ is again truncated after the constant term, for simplicity, i.e. $\boldsymbol{H}(\boldsymbol{x}') \approx \boldsymbol{H}_1(\boldsymbol{x}') = \boldsymbol{H}(\boldsymbol{x})$. The influence of the linear in-plane gradient $\nabla_{\mathrm{p}}\boldsymbol{H}$ will subsequently be discussed. In order to compute the microscopic displacement gradient $\boldsymbol{H}^*(\boldsymbol{x}')$ the fluctuation gradient is derived from Eq. (3.74)

$$\begin{aligned} \boldsymbol{H}^*_{12}(\boldsymbol{x}') &= \boldsymbol{H}_1(\boldsymbol{x}') + \tilde{\boldsymbol{H}}_2(\boldsymbol{x}') \\ &= \boldsymbol{H}(\boldsymbol{x}) - \gamma_2(\boldsymbol{x}^{\mathrm{P}}(\boldsymbol{x}'))\boldsymbol{d}\otimes\boldsymbol{n} - \Delta z(\boldsymbol{x}')\boldsymbol{d}\otimes\nabla_{\mathrm{p}}\gamma(\boldsymbol{x}). \end{aligned} \tag{3.77}$$

Note that the components which have been identified with bending-like micro-deformations $(\boldsymbol{d}\cdot\nabla\gamma)$ and torsion-like micro-deformations $(\boldsymbol{h}\cdot\nabla\gamma)$ in Sect. 3.2 also appear here (see Eq. (3.73)). Before computing the energy $W_{12}(\boldsymbol{x})$, it is worthwhile to investigate the stresses in the averaging volume $\Delta V(\boldsymbol{x})$

$$\begin{aligned} \boldsymbol{\sigma}^*_{12}(\boldsymbol{x}') &= \mathbb{C}[\boldsymbol{H}^*_{12}(\boldsymbol{x}')] = \mathbb{C}[\boldsymbol{H}_1(\boldsymbol{x}') + \tilde{\boldsymbol{H}}_2(\boldsymbol{x}')] \\ &\overset{(3.77)}{=} \mathbb{C}[\boldsymbol{H}_1(\boldsymbol{x}) - \gamma_2(\boldsymbol{x}^{\mathrm{P}}(\boldsymbol{x}'))\boldsymbol{d}\otimes\boldsymbol{n}] - \Delta z(\boldsymbol{x}')\mathbb{C}[\boldsymbol{d}\otimes\nabla_{\mathrm{p}}\gamma(\boldsymbol{x})]. \end{aligned} \tag{3.78}$$

If the stresses are evaluated over the height of a given layer (i.e. along a line segment with direction $\boldsymbol{n}$), two contributions can be identified: a constant term $\mathbb{C}[\boldsymbol{H}_1(\boldsymbol{x}) - \gamma_2(\boldsymbol{x}^{\mathrm{P}}(\boldsymbol{x}'))\boldsymbol{d}\otimes\boldsymbol{n}]$ and a fluctuation term which depends linearly on the distance $\Delta z(\boldsymbol{x}')$ from the layer center plane $\mathcal{L}_i$. The latter term is given by

$$-\Delta z(\boldsymbol{x}')\mathbb{C}[\boldsymbol{d}\otimes\nabla_{\mathrm{p}}\gamma(\boldsymbol{x})] = -\Delta z(\boldsymbol{x}')(\boldsymbol{h}\cdot\nabla\gamma(\boldsymbol{x}))\,\mathbb{C}[\boldsymbol{d}\otimes\boldsymbol{h}] \tag{3.79}$$

$$\qquad - \Delta z(\boldsymbol{x}')(\boldsymbol{d}\cdot\nabla\gamma(\boldsymbol{x}))\,\mathbb{C}[\boldsymbol{d}\otimes\boldsymbol{d}]. \tag{3.80}$$

For the special case of an isotropic material the first term of the right hand side represents shear stresses in the $\boldsymbol{h}$-$\boldsymbol{d}$-plane (with respect to the $\{\boldsymbol{h}, \boldsymbol{d}, \boldsymbol{n}\}$-coordinate system)

$$\tau_{\mathrm{t}} = -\Delta z(\boldsymbol{x}')\, G\, (\boldsymbol{h} \cdot \nabla\gamma(\boldsymbol{x})). \tag{3.81}$$

The second term represents normal stresses. For example, the component in $\boldsymbol{d}$-direction is given by

$$\sigma_{\mathrm{b}} = -\Delta z(\boldsymbol{x}')(\boldsymbol{d} \cdot \nabla\gamma(\boldsymbol{x}))\, \frac{E}{1+\upsilon}\frac{1-\upsilon}{1-2\upsilon}. \tag{3.82}$$

These stresses are illustrated in Fig. 3.12.

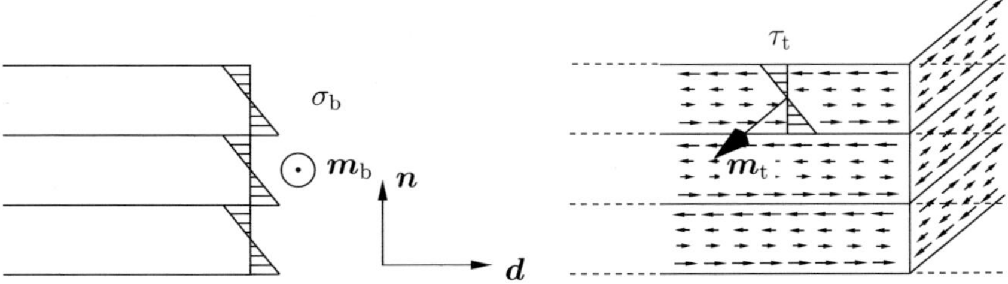

Figure 3.12: Visualization of the stresses $\sigma_{\mathrm{b}}(\Delta z)$ and $\tau_{\mathrm{t}}(\Delta z)$.

Clearly, σ_{b} reminds of bending stresses and τ_{t} can be associated with torsion. The stresses are proportional to the components $\boldsymbol{d} \cdot \nabla\gamma$ and $\boldsymbol{h} \cdot \nabla\gamma$, respectively (compare Sect. 3.2). In fact, Fig. 3.12 could identically have appeared in Sect. 3.2 illustrating the bending and torsion stresses in the layers. The major difference, compared to Sect. 3.2, is the consideration of lateral stresses (Eqns. (3.80) and (3.82)). These have been neglected in the bending example in Sect. 3.2. Consistency is achieved in the special case $\upsilon \to 0$. In that case, the bending stress reads $\sigma_{\mathrm{b}} = -\Delta z(\boldsymbol{x}')E(\boldsymbol{d} \cdot \nabla\gamma(\boldsymbol{x}))$ and all other normal stresses vanish.

In the following, the considerations are restricted to (anisotropic) cubic crystals. In the subsequent sections, the moments generated by σ_{b} and τ_{t} will reappear. The bending moment is given by

$$m_{\mathrm{b}}(\boldsymbol{x}) = -\frac{1}{h}\int\limits_{-h/2}^{h/2} \Delta z\, \sigma_{\mathrm{b}}\, \mathrm{d}\Delta z. \tag{3.83}$$

More precisely, m_b represents an average moment per unit area. Note that $m_\mathrm{b}(\boldsymbol{x})$ could have been computed anywhere in the averaging volume $\Delta V(\boldsymbol{x})$. The evaluation of the integral yields

$$m_\mathrm{b}(\boldsymbol{x}) = \frac{h^2}{12} C^\mathrm{dd} \big(\boldsymbol{d} \cdot \nabla \gamma(\boldsymbol{x})\big), \tag{3.84}$$

with $C^\mathrm{dd} = (\boldsymbol{d} \otimes \boldsymbol{d}) \cdot \mathbb{C}[\boldsymbol{d} \otimes \boldsymbol{d}]$.
Similarly, the torsion moment per unit area is given by

$$m_\mathrm{t}(\boldsymbol{x}) = -\frac{1}{h} \int_{-h/2}^{h/2} \Delta z \, \tau_\mathrm{t} \, \mathrm{d}\Delta z = \frac{h^2}{12} C^\mathrm{dh} \big(\boldsymbol{h} \cdot \nabla \gamma(\boldsymbol{x})\big), \tag{3.85}$$

with $C^\mathrm{dh} = (\boldsymbol{d} \otimes \boldsymbol{h}) \cdot \mathbb{C}[\boldsymbol{d} \otimes \boldsymbol{h}]$.
In general, moments are vectors, i.e., they have a direction. Figure 3.12 motivates the following directions of m_b and m_t

$$\boldsymbol{m}_\mathrm{b}(\boldsymbol{x}) = m_\mathrm{b}(\boldsymbol{x}) \boldsymbol{h}, \quad \boldsymbol{m}_\mathrm{t}(\boldsymbol{x}) = m_\mathrm{t}(\boldsymbol{x}) \boldsymbol{h}. \tag{3.86}$$

The shear stresses $\tau_\mathrm{t} \boldsymbol{d} \otimes \boldsymbol{h}$ imply shear stresses $\tau_\mathrm{t} \boldsymbol{h} \otimes \boldsymbol{d}$ due to the symmetry of the stress tensor (see Fig. 3.12). The associated moment reads

$$\boldsymbol{m}_\mathrm{t}^\perp(\boldsymbol{x}) = -m_\mathrm{t}(\boldsymbol{x}) \boldsymbol{d}. \tag{3.87}$$

Finally, the energy $W_{12}(\boldsymbol{x})$ can be computed based on approximation (3.77) of $\boldsymbol{H}^*(\boldsymbol{x}')$

$$W_{12}(\boldsymbol{x}) = \lim_{\Delta V(\boldsymbol{x}) \to 0} \frac{1}{\Delta V(\boldsymbol{x})} \int_{\Delta V(\boldsymbol{x})} \frac{1}{2} \boldsymbol{H}_{12}^*(\boldsymbol{x}') \cdot \mathbb{C}[\boldsymbol{H}_{12}^*(\boldsymbol{x}')] \, \mathrm{d}v'. \tag{3.88}$$

It can be shown (App. B.4) that

$$W_{12}(\boldsymbol{x}) = \frac{1}{2} \boldsymbol{\varepsilon}^\mathrm{e}(\boldsymbol{x}) \cdot \mathbb{C}[\boldsymbol{\varepsilon}^\mathrm{e}(\boldsymbol{x})] + \frac{1}{2} \frac{h^2}{12} (\boldsymbol{d} \otimes \nabla_\mathrm{p} \gamma(\boldsymbol{x})) \cdot \mathbb{C}[\boldsymbol{d} \otimes \nabla_\mathrm{p} \gamma(\boldsymbol{x})] \tag{3.89}$$

$$= \frac{1}{2} \boldsymbol{\varepsilon}^\mathrm{e}(\boldsymbol{x}) \cdot \mathbb{C}[\boldsymbol{\varepsilon}^\mathrm{e}(\boldsymbol{x})] + \frac{1}{2} \frac{h^2}{12} \big(C^\mathrm{dd} (\boldsymbol{d} \cdot \nabla \gamma(\boldsymbol{x}))^2 + C^\mathrm{dh} (\boldsymbol{h} \cdot \nabla \gamma(\boldsymbol{x}))^2\big). \tag{3.90}$$

In the special case of elastic isotropy and $\upsilon \to 0$, one obtains the result (3.32), which has been guessed based on the fundamental bending and torsion examples.

Remarks:

- The stress fluctuations vanish upon averaging of $\boldsymbol{\sigma}^*_{12}(\boldsymbol{x}')$ (Eq. (3.78)) over $\Delta V(\boldsymbol{x})$. The remaining average stress reads

$$\bar{\boldsymbol{\sigma}}_{12}(\boldsymbol{x}) = \frac{1}{\Delta V(\boldsymbol{x})} \int_{\Delta V(\boldsymbol{x})} \boldsymbol{\sigma}^*_{12}(\boldsymbol{x}')\,\mathrm{d}v' = \mathbb{C}[\boldsymbol{\varepsilon}(\boldsymbol{x}) - \gamma(\boldsymbol{x})\boldsymbol{P}^{\mathrm{S}}] = \mathbb{C}[\boldsymbol{\varepsilon}^{\mathrm{e}}(\boldsymbol{x})], \tag{3.91}$$

  with $\boldsymbol{P}^{\mathrm{S}} = \mathrm{sym}(\boldsymbol{d} \otimes \boldsymbol{n})$.
- Formally, the energy (3.89) is close to the energy (1.28) proposed by Gurtin et al. (2007), where it is postulated based on dislocation arguments.
- In contrast to Sect. 3.2, the Bernoulli-Hypothesis is in general not fulfilled in the approach at hand. This means that the microstresses depicted in Fig. 3.12 do not explicitly depend on the curvature of the layers. Therefore, the present theory might rather be compared to the Timoshenko beam than to the Bernoulli theory (as far as bending is concerned). The normal stresses in the Timoshenko beam can take the typical linear shape (Fig. 3.12, left) even if the beam is remains straight. This is a result of the decoupling of the cross section rotation from the slope of the bending curve which constitutes the basic distinction between the Timoshenko and the Bernoulli theory.
- The fluctuation-approximation associated to bending involves tensile and compressive regions (Fig. 3.12). These stresses cause lateral contractions which are not yet accounted for. In order to compute more realistic energy approximations, these lateral contractions should be taken into account, additionally (this is not within the scope of the present work).
- As a next step, it seems natural to enrich the formulation by taking into account the influence of $\nabla_{\mathrm{p}}\boldsymbol{H}$, in addition (in Eq. (3.77)). This would lead to the further improved energy W_{22}. However, some calculations show that this does not alter the result (3.89), i.e. $W_{22} = W_{12}$. The proof follows the same lines as App. B.4.

- The dimensions of the averaging volume are significantly larger in the in-plane directions than in the out-of-plane direction. This special choice simplifies the derivation and takes into accout that the plastic slip may become non-smooth in the direction of the slip plane normal (Fig. 3.7). However, the chosen shape of the averaging volume may limit the accuracy of the derived energies. An evaluation of this point may lead to further insights, but is beyond the scope of this work.
- Yet, the slip band distance h has been considered a constant. However, in a generalized setting it may be assumed that h changes during the plastic deformation process. For example, one may assume that only few slip bands are present at the beginning of the deformation and that additional ones occur with ongoing deformation.

  In a first rough approximation, this effect may be accounted for as follows. Since the dislocation density also tends to increase with ongoing deformation and increasing number of slip bands (more slip bands "carry" more dislocations), the slip band distance h may be approximated as a function of the dislocation density. If, in addition, the loading is monotonous and the problem is GND-dominated, the total dislocation density may be approximated by the GND-density. For example, if exclusively edge dislocations are present, the slip band distance could then be estimated by a function $h(|\rho_{\vdash}|)$, as illustrated in Fig. 3.13.

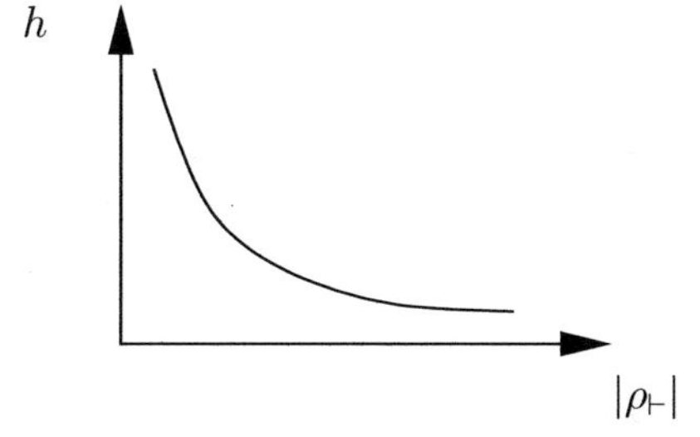

Figure 3.13: Slip band distance as a function of the GND-density.

In order to exemplarily investigate the effect of such an approach on the energy, consider, for example, the ansatz $h = \sqrt{c_h b/|\rho_\vdash|}$, where c_h is a dimensionless constant. This ansatz implies the assumption that the average dislocation distance inside the individual slip bands scales like the dislocation distance normal to the slip bands (i.e. the slip band spacing). In this case, the plastic strain gradient-depended part of the energy (3.89) reduces to

$$W_{\mathrm{g}} = \frac{c_h\, b\, C^{\mathrm{dd}}}{24} |\rho_\vdash|. \tag{3.92}$$

Hence, in this simplified setting, the energy represents a rank-one potential, as discussed in the context of a laminate material in Chap. 2 (Eq. (2.30)) and used, e.g., by Ortiz and Repetto (1999).

3.4 Dissipation Potential

The relative displacement in the slip bands is assumed to be caused by dislocation motion. If no external load is applied the dislocations are assumed to be trapped in energetically favorable configurations. Additionally, their motion is hindered due to obstacles. Therefore, pronounced dislocation motion is expected to occur only if the local shear stress exceeds a critical value.
Introducing the time as additional coordinate, this behavior is modeled phenomenologically by a dissipation potential $\phi_v(v_{\mathrm{rel}})$, where $v_{\mathrm{rel}} = \boldsymbol{d} \cdot [\![\dot{\boldsymbol{u}}]\!]$ denotes the relative velocity of adjacent layers in the slip bands. The dissipation per unit area (of the slip band) is given by

$$\partial_{v_{\mathrm{rel}}} \phi_v(v_{\mathrm{rel}}) v_{\mathrm{rel}}, \tag{3.93}$$

i.e., for $v_{\mathrm{rel}} \neq 0$ the expression $\tau^{\mathrm{d}} = \partial_{v_{\mathrm{rel}}} \phi_v(v_{\mathrm{rel}})$ represents a dissipative shear stress. In order to account for a possible elastic range, the symbol ∂ may be understood as a sub-differential operator (see, e.g., Han and

Reddy, 2013, and Sect. 2.1). The plastic shear rate is given by (cf. Eq. (3.40))

$$\dot{\gamma}(\boldsymbol{x},t) = \frac{1}{\Delta V(\boldsymbol{x})} \int\limits_{\Delta V(\boldsymbol{x})} \partial_t j_{\mathrm{u}}(\boldsymbol{x}',t)\, \delta_{\mathcal{SB}}(\boldsymbol{x}')\, \mathrm{d}v'. \tag{3.94}$$

It is assumed that $j_{\mathrm{u}}(\boldsymbol{x}',t)$ can be approximated by the following linear approach inside of $\Delta V(\boldsymbol{x})$

$$j_{\mathrm{u}}(\boldsymbol{x}',t) \approx j_{\mathrm{u},2}(\boldsymbol{x}',t) = j_{\mathrm{u}0}(\boldsymbol{x},t) + \boldsymbol{b}_0(\boldsymbol{x},t) \cdot (\boldsymbol{x}' - \boldsymbol{x}). \tag{3.95}$$

In the slip bands, the following relation holds

$$v_{\mathrm{rel}}(\boldsymbol{x},t) = \partial_t j_{\mathrm{u}}(\boldsymbol{x},t), \quad \boldsymbol{x} \in \mathcal{SB}. \tag{3.96}$$

The insertion of Eq. (3.95) into Eq. (3.94) yields

$$\dot{\gamma}(\boldsymbol{x},t) \approx \frac{1}{h} \partial_t\, j_{\mathrm{u}0}(\boldsymbol{x},t) = \frac{1}{h} v_{\mathrm{rel}}(\boldsymbol{x},t). \tag{3.97}$$

This result leads to following integral dissipation potential approximation

$$\sum_i \int\limits_{\mathcal{SB}_i} \phi_v(v_{\mathrm{rel}}(\boldsymbol{x}))\, \mathrm{d}a \approx \int\limits_{\mathcal{B}} \frac{1}{h} \phi_v(h\, \dot{\gamma}(\boldsymbol{x}))\, \mathrm{d}v. \tag{3.98}$$

The latter integral motivates the definition of the dissipation potential in terms of the plastic slip rate

$$\phi(\dot{\gamma}) = \frac{1}{h} \phi_v(h\, \dot{\gamma}). \tag{3.99}$$

Remarks:

- In the rate independent case, the dissipation potential is homogeneous of degree one and consequently

$$\phi(\dot{\gamma}) = \phi_v(\dot{\gamma}). \tag{3.100}$$

- The rate dependent case may, e.g., be modeled by a dissipation potential of the power law type, which is homogeneous of degree $p+1 > 1$

$$\phi(\omega\dot{\gamma}) = \omega^{(p+1)} \phi(\dot{\gamma}), \quad \omega \in \mathbb{R}. \tag{3.101}$$

This kind of potential is consistent with a potential $\phi_v(v_{\text{rel}})$ which is homogeneous of degree $p+1$ since in that case

$$\phi(\omega\dot{\gamma}) = \frac{1}{h}\phi_v(\omega h\,\dot{\gamma}) = \omega^{(p+1)}\frac{1}{h}\phi_v(h\,\dot{\gamma}) = \omega^{(p+1)}\phi(\dot{\gamma}). \tag{3.102}$$

- Eq. (3.99) implies that rate dependent dissipation potentials ϕ depend on the internal length scale h. For example, a typical power law approach is considered

$$\phi_v(v_{\text{rel}}) = \frac{p\,\tau^{\text{D}}\,v_0}{(p+1)}\left(\frac{|v_{\text{rel}}|}{v_0}\right)^{1+\frac{1}{p}}. \tag{3.103}$$

Here, the drag stress τ^{D}, the rate sensitivity p and the reference velocity v_0 are phenomenological material parameters. The dissipative shear stress is given by

$$\tau^{\text{d}} = \partial_{v_{\text{rel}}}\phi_v = \partial_{\dot{\gamma}}\phi = \operatorname{sgn}(\dot{\gamma})\,\tau^{\text{D}}\left(\frac{|\dot{\gamma}|}{v_0/h}\right)^{\frac{1}{p}}. \tag{3.104}$$

The denominator v_0/h can be interpreted as reference shear rate $\dot{\gamma}_0$ which is the lower the smaller the distance h between the slip bands.

3.5 A Gradient Plasticity Theory based on W_{12}

The principle of virtual power is exploited in order to derive field equations for cubic crystals. A single crystal $\mathcal{B}$ in a single slip situation is considered. Since the mesoscopic scale is the only length scale considered in this section, the following notations are used. The strain tensor $\boldsymbol{\varepsilon} = \text{sym}(\boldsymbol{H})$ is the symmetric part of the displacement gradient $\boldsymbol{H} = \nabla\boldsymbol{u}$. The plastic strain (in terms of the plastic shear strain γ) reads $\boldsymbol{\varepsilon}^{\text{p}} = \gamma\boldsymbol{P}^{\text{S}}$ with $\boldsymbol{P}^{\text{S}} = \text{sym}(\boldsymbol{d}\otimes\boldsymbol{n})$. Here, $\boldsymbol{d}$ denotes the slip direction and $\boldsymbol{n}$ is the slip plane normal. The vector $\boldsymbol{h} = \boldsymbol{d}\times\boldsymbol{n}$ allows the definition of the orthonormal system $\{\boldsymbol{h}, \boldsymbol{d}, \boldsymbol{n}\}$.

The theory is based on the following stored energy approach (cf. Sect. 3.3.2, Eq. (3.89))

$$W = W_{\mathrm{e}}(\boldsymbol{\varepsilon}, \gamma) + W_{\mathrm{g}}(\nabla\gamma) \tag{3.105}$$

with the elastic energy

$$W_{\mathrm{e}} = \frac{1}{2}\boldsymbol{\varepsilon}^{\mathrm{e}} \cdot \mathbb{C}[\boldsymbol{\varepsilon}^{\mathrm{e}}] = \frac{1}{2}(\boldsymbol{\varepsilon} - \gamma\boldsymbol{P}^{\mathrm{S}}) \cdot \mathbb{C}[\boldsymbol{\varepsilon} - \gamma\boldsymbol{P}^{\mathrm{S}}] \tag{3.106}$$

and the gradient extension

$$W_{\mathrm{g}} = \frac{1}{2}\frac{h^2}{12}\left(C^{\mathrm{dd}}(\boldsymbol{d}\cdot\nabla\gamma)^2 + C^{\mathrm{dh}}(\boldsymbol{h}\cdot\nabla\gamma)^2\right) \tag{3.107}$$

$$= \frac{1}{2}\left(m_{\mathrm{b}}\,(\boldsymbol{d}\cdot\nabla\gamma) + m_{\mathrm{t}}\,(\boldsymbol{h}\cdot\nabla\gamma)\right) \tag{3.108}$$

with $C^{\mathrm{dd}} = (\boldsymbol{d}\otimes\boldsymbol{d})\cdot\mathbb{C}[\boldsymbol{d}\otimes\boldsymbol{d}]$ and $C^{\mathrm{dh}} = (\boldsymbol{d}\otimes\boldsymbol{h})\cdot\mathbb{C}[\boldsymbol{d}\otimes\boldsymbol{h}]$ (cf. Eqns. (3.84) and (3.85)). Here, the slip band distance h is considered a constant, for simplicity. The microscopic moments m_{b} and m_{t} are given by

$$m_{\mathrm{b}} = \frac{h^2}{12}C^{\mathrm{dd}}(\boldsymbol{d}\cdot\nabla\gamma),\; m_{\mathrm{t}} = \frac{h^2}{12}C^{\mathrm{dh}}(\boldsymbol{h}\cdot\nabla\gamma). \tag{3.109}$$

They are the components of the energetical micro-moment vector

$$\boldsymbol{m} = \partial_{\nabla\gamma}W_{\mathrm{g}} = m_{\mathrm{b}}\boldsymbol{d} + m_{\mathrm{t}}\boldsymbol{h}, \tag{3.110}$$

which is formally identical to the (energetic) microstresses $\boldsymbol{\xi}_\alpha$ (1.26) of Gurtin's gradient plasticity theory. The slip band distance h represents the internal length scale of the model. Note that for single slip and a properly chosen quadratic defect energy W_{g}, the generalized stresses $\boldsymbol{m}$ and $\boldsymbol{\xi}$ are fully identical. For simplicity, ansatz (3.105) does not account for isotropic hardening.

Moreover, the inelastic behavior is modeled by a dissipation potential $\phi = \phi(\dot{\gamma})$ (cf. Sect. 3.4). As a consequence, the following internal forces can be identified

$$\boldsymbol{\sigma} = \partial_{\varepsilon}W_{\mathrm{e}}, \quad \tau = -\partial_{\gamma}W_{\mathrm{e}}, \quad \boldsymbol{m} = \partial_{\nabla\gamma}W_{\mathrm{g}}, \quad \tau^{\mathrm{d}} = \partial_{\dot{\gamma}}\phi. \tag{3.111}$$

The virtual power of the internal forces is assumed to be given by

$$\delta\mathcal{P}_{\text{int}}(\Delta\mathcal{B}) = \int_{\Delta\mathcal{B}} (\delta\dot{W} + \partial_{\dot{\gamma}}\phi\,\delta\dot{\gamma})\,\mathrm{d}v. \tag{3.112}$$

The set $\Delta\mathcal{B} \subseteq \mathcal{B}$ is an arbitrary subset of the body $\mathcal{B}$. The virtual power $\delta\dot{W}$ of the energetical stresses is given by

$$\begin{aligned} \delta\dot{W} &= \partial_{\varepsilon}W_{\mathrm{e}} \cdot \delta\dot{\varepsilon} + \partial_{\gamma}W_{\mathrm{e}}\,\delta\dot{\gamma} + \partial_{\nabla\gamma}W_{\mathrm{g}} \cdot \nabla(\delta\dot{\gamma}) && (3.113)\\ &= \boldsymbol{\sigma} \cdot \delta\dot{\boldsymbol{\varepsilon}} - \tau\,\delta\dot{\gamma} + \boldsymbol{m} \cdot \nabla(\delta\dot{\gamma}). && (3.114) \end{aligned}$$

Here, $\delta\dot{\boldsymbol{u}}$ denotes the virtual velocity, $\delta\dot{\gamma}$ is a virtual plastic shear rate and $\delta\dot{\boldsymbol{\varepsilon}} = \mathrm{sym}(\nabla\delta\dot{\boldsymbol{u}})$ is the virtual strain rate. The application of the chain rule and Gauss' theorem yields

$$\begin{aligned} \delta\mathcal{P}_{\text{int}}(\Delta\mathcal{B}) &= \int_{\Delta\mathcal{B}} \big(\boldsymbol{\sigma} \cdot \delta\dot{\boldsymbol{\varepsilon}} - (\tau - \tau^{\mathrm{d}})\,\delta\dot{\gamma} + \boldsymbol{m} \cdot \nabla(\delta\dot{\gamma})\big)\,\mathrm{d}v && (3.115)\\ &= -\int_{\Delta\mathcal{B}} \big(\mathrm{div}\,(\boldsymbol{\sigma}) \cdot \delta\dot{\boldsymbol{u}} + (\tau + \mathrm{div}\,(\boldsymbol{m}) - \tau^{\mathrm{d}})\,\delta\dot{\gamma}\big)\,\mathrm{d}v \\ &\quad + \int_{\partial\Delta\mathcal{B}} \big((\boldsymbol{\sigma}\tilde{\boldsymbol{n}}) \cdot \delta\dot{\boldsymbol{u}} + (\boldsymbol{m} \cdot \tilde{\boldsymbol{n}})\,\delta\dot{\gamma}\big)\,\mathrm{d}a, \end{aligned}$$

where $\tilde{\boldsymbol{n}}$ is the outer normal. The virtual power of the external forces is assumed to be given by

$$\delta\mathcal{P}_{\text{ext}}(\Delta\mathcal{B}) = \int_{\partial\Delta\mathcal{B}} \big(\boldsymbol{t} \cdot \delta\dot{\boldsymbol{u}} + m_{\mathrm{a}}\,\delta\dot{\gamma}\big)\,\mathrm{d}a. \tag{3.116}$$

Here, $\boldsymbol{t}$ denotes the traction vector and m_{a} is a micro moment. The principle of virtual power postulates the equality of $\delta\mathcal{P}_{\text{ext}}(\Delta\mathcal{B})$ and $\delta\mathcal{P}_{\text{int}}(\Delta\mathcal{B})$ for arbitrary virtual fields. This implies various equations. The first one reads

$$\boldsymbol{t} = \boldsymbol{\sigma}\tilde{\boldsymbol{n}}. \tag{3.117}$$

The second one states that the externally applied micro moment m_{a} partially equilibrates the two internal micro moments

$$m_{\mathrm{a}} = \boldsymbol{m} \cdot \tilde{\boldsymbol{n}} = m_{\mathrm{b}}(\boldsymbol{d} \cdot \tilde{\boldsymbol{n}}) + m_{\mathrm{t}}(\boldsymbol{h} \cdot \tilde{\boldsymbol{n}}). \tag{3.118}$$

Moreover, the classical linear momentum balance is preserved

$$\operatorname{div}(\boldsymbol{\sigma}) = \mathbf{0}, \tag{3.119}$$

and a micro moment equation is obtained

$$\tau + \operatorname{div}(\boldsymbol{m}) - \tau^{\mathrm{d}} = 0. \tag{3.120}$$

These equations are formally close to the equations obtained, e.g., by Gurtin et al. (2007).
In general, the boundary conditions are given as follows: the displacement is prescribed on a subset $\partial\mathcal{B}_u \subseteq \partial\mathcal{B}$ by a given function $\bar{\boldsymbol{u}}(\boldsymbol{x}, t)$. The tractions are prescribed by the function $\bar{\boldsymbol{t}}(\boldsymbol{x}, t)$ on the remaining part $\partial\mathcal{B}_t = \partial\mathcal{B} \setminus \partial\mathcal{B}_u$. Similarly, the plastic shear strain is prescribed on $\partial\mathcal{B}_\gamma \subseteq \partial\mathcal{B}$ by the function $\bar{\gamma}(\boldsymbol{x}, t)$ and the micro moment is prescribed by $\bar{m}_{\mathrm{a}}(\boldsymbol{x}, t)$ on $\partial\mathcal{B}_m = \partial\mathcal{B} \setminus \partial\mathcal{B}_\gamma$.
Eq. (3.120) represents the continuum generalization of the moment equation which is well known from the classical beam theory. This physical interpretation will be further investigated in the next section. Moreover, Eq. (3.120) is formally identical to the microforce balance equation (Eq. (1.27)) of Gurtin's gradient plasticity theory.

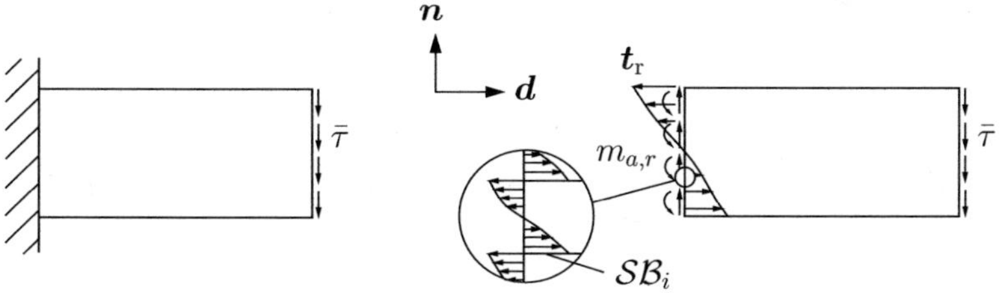

Figure 3.14: Visualization (qualitative) of the stress vector $\boldsymbol{t}_{\mathrm{r}}$ (support reactions) and the microscopic representation of the micro reaction moment $m_{\mathrm{a,r}}$. Here $m_{\mathrm{a,r}} = \boldsymbol{m} \cdot \tilde{\boldsymbol{n}} = -m_{\mathrm{b}}$.

In order to physically interpret Eq. (3.118) a simple example is considered (see Fig. 3.14). An elastoplastic cantilever beam is loaded by a prescribed shear stress $\bar{\boldsymbol{t}} = \bar{\tau}\boldsymbol{e}_z$ on its right end. The slip plane normal and slip direction are illustrated in Fig. 3.14. The load is assumed sufficiently large, such that the beam deformation is fully elasto-plastic. The support reactions at the left end are given by the reaction stress vector $\boldsymbol{t}_\mathrm{r}$ and distributed reaction micro moment $m_{\mathrm{a,r}}$. The expected support reactions are qualitatively illustrated in Fig. 3.14. In this example, the normal vector at the support is given by $\tilde{\boldsymbol{n}} = -\boldsymbol{d}$. As a consequence, the reaction micro moment density $m_{\mathrm{a,r}}$ reads

$$m_{\mathrm{a,r}} = \boldsymbol{m} \cdot \tilde{\boldsymbol{n}} = (m_\mathrm{b}\, \boldsymbol{d} + m_\mathrm{t}\, \boldsymbol{h}) \cdot (-\boldsymbol{d}) = -m_\mathrm{b}. \qquad (3.121)$$

3.6 Polycrystals and Grain Boundaries

3.6.1 Beam Theory Solution

A simplified polycrystal model, as illustrated in Fig. 3.15, is considered in plane strain in order to investigate systems consisting of several grains.

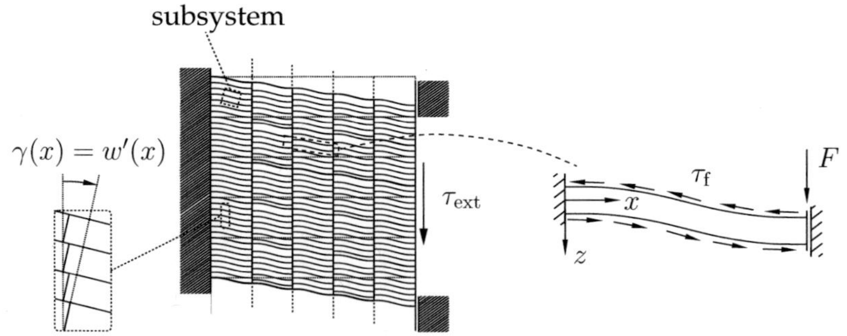

Figure 3.15: Simplified polycrystal model.

Each grain is modeled by a cube with edge length D_g consisting of $N_\mathrm{g} \gg 1$ layers with thickness $h = D_\mathrm{g}/N_\mathrm{g}$. On the left, the model is supported by a built-in support. Its right end can be displaced vertically and is loaded by an external shear stress τ_ext. In between the layers, a shear stress τ_f with friction character is assumed. For simplicity, each layer is assumed to behave like a beam which is loaded by a force $F = \tau_{\mathrm{ext}} D_{\mathrm{g}} h$ (Fig. 3.15, right).

The slip bands of different layers are assumed incompatible. This means that each slip band ends at the grain boundary since the vertical slip band positions of adjacent grains are not matching. As a consequence, the rotation of the beam ends is constrained to vanish. The grain boundaries can therefore be considered impenetrable for dislocations. For the special case $\tau_{\mathrm{f}} = 0$, the bending curve can be computed from the differential equation of the bending line

$$w''(x) = -\frac{M_{\mathrm{b}}(x)}{EI} \tag{3.122}$$

and the boundary conditions $w(0) = w'(0) = w'(D_{\mathrm{g}}) = 0$. Here, M_{b} denotes the bending moment and shear influences have been neglected. The solution of Eq. (3.122) reads

$$w(x) = \frac{\tau_{\mathrm{ext}}}{Eh^2}(3D_{\mathrm{g}}x^2 - 2x^3). \tag{3.123}$$

The derivative of $w(x)$ can be interpreted as average plastic shear strain $\gamma(x) = w'(x)$ (cf. Fig. 3.15, left).

As already mentioned, shear force influences have been neglected since the Bernoulli beam theory has been applied. The shear influence can be approximately accounted for by assuming a homogeneous shear deformation $\varepsilon_{xz} = \tau_{\mathrm{ext}}/(2G)$ of the beam. The superposition of the associated displacement with the bending deflection (3.123) reads

$$w(x) = \frac{\tau_{\mathrm{ext}}}{Eh^2}(3D_{\mathrm{g}}x^2 - 2x^3) + \frac{\tau_{\mathrm{ext}}}{G}x. \tag{3.124}$$

In order to interpret the equilibrium equation (3.120) a cuboid shaped subsystem of a grain with edge lengths Δx, $\Delta y = D_{\mathrm{g}}$ and Δz is considered (see Fig. 3.15, top). The subsystem consists of $n = \Delta z/h$ layers.

In Fig. 3.16, the free body diagram of the subsystem is illustrated. The associated equilibrium of moments reads

$$-nM_{\mathrm{b}}(x) + nM_{\mathrm{b}}(x + \Delta x) - nQ(x)\Delta x + \overline{\tau}_{\mathrm{f}}\Delta x \Delta y \Delta z = 0, \tag{3.125}$$

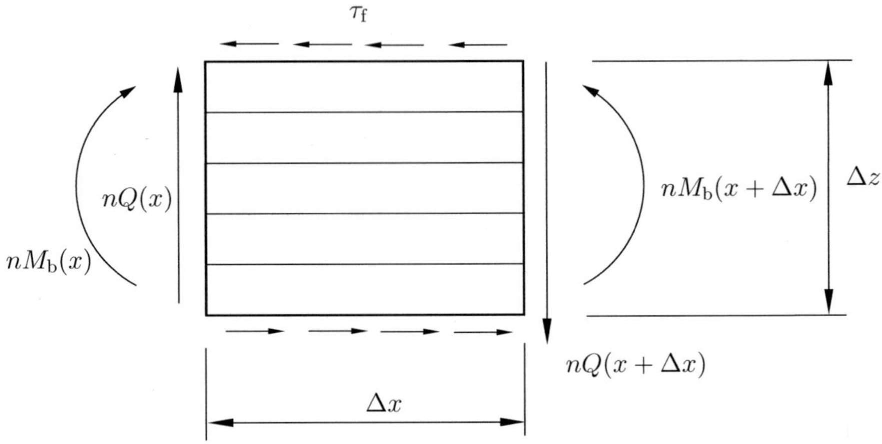

Figure 3.16: Free body diagram of the subsystem as indicated in Fig. 3.15.

with

$$\overline{\tau}_f = \frac{1}{\Delta x} \int\limits_{x}^{x+\Delta x} \tau_f(\bar{x})\,d\bar{x}. \tag{3.126}$$

Introducing the average shear stress and average bending moment per unit area

$$\overline{\tau}(x) = \frac{nQ(x)}{\Delta y \Delta z}, \quad \overline{m}_b(x) = -\frac{nM_b(x)}{\Delta y \Delta z} \tag{3.127}$$

and division of Eq. (3.125) by $-\Delta x \Delta y \Delta z$ yields

$$\frac{\overline{m}_b(x+\Delta x) - \overline{m}_b(x)}{\Delta x} + \overline{\tau}(x) - \overline{\tau}_f(x) = 0. \tag{3.128}$$

In the limit $\Delta x \to 0$ the following identity is obtained

$$\overline{\tau} + \operatorname{div}(\overline{m}_b \boldsymbol{d}) - \overline{\tau}_f = 0. \tag{3.129}$$

This equation is formally identical to Eq. (3.120), given by

$$\tau + \operatorname{div}(\boldsymbol{m}) - \tau^d = 0. \tag{3.130}$$

Both equations describe almost the same mechanics. The second equation also accounts for torsion like moments.

3.6.2 Continuum Theory Solution

Subsequently, the continuum theory derived in Sect. 3.5 is applied to solve the problem considered in the previous section. For simplicity, only one crystal is considered (see Fig. 3.17). The material is assumed to be elastically isotropic with $\upsilon \to 0$. The displacement and the plastic shear strain are assumed to be functions of x

$$\boldsymbol{u} = w(x)\boldsymbol{e}_z, \quad \gamma = \gamma(x). \tag{3.131}$$

In addition, the boundary conditions are given by

$$w(0) = 0, \quad \gamma(0) = \gamma(D_{\mathrm{g}}) = 0, \quad \sigma_{xz}(x = D_{\mathrm{g}}) = \sigma_{zx}(x = D_{\mathrm{g}}) = \tau_{\mathrm{ext}}. \tag{3.132}$$

The conditions $(3.132)_2$ represent the slip band incompatibility at the grain boundaries. Moreover, the gradient of γ reads $(\boldsymbol{d} \cdot \nabla\gamma)\boldsymbol{d} = \gamma'(x)\boldsymbol{d}$

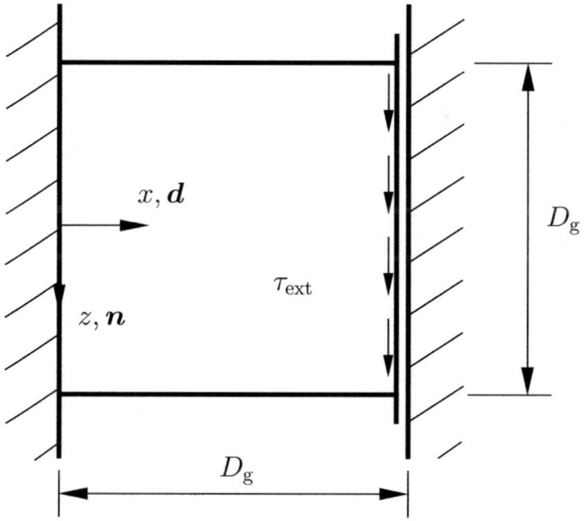

Figure 3.17: Crystal loaded by constant shear stress τ_{ext}.

and the constant C^{dd} is given by $C^{\mathrm{dd}} = E$, since $\upsilon \to 0$. As a consequence, the stored energy (3.105) is given by

$$W = \frac{1}{2}\boldsymbol{\varepsilon}^{\mathrm{e}} \cdot \mathbb{C}[\boldsymbol{\varepsilon}^{\mathrm{e}}] + \frac{1}{2}\frac{h^2}{12}E\big(\gamma'(x)\big)^2. \tag{3.133}$$

The micro moment reads

$$\boldsymbol{m} = \partial_{\nabla\gamma} W = \frac{h^2}{12} E\gamma'(x)\boldsymbol{d}. \tag{3.134}$$

Since friction between the layers has been neglected in the previous section, the dissipation potential ϕ is analogously set to zero. The elastic strain can be derived from $\boldsymbol{H}$ and $\boldsymbol{H}^{\mathrm{p}}$

$$\boldsymbol{H} = w'(x)\boldsymbol{e}_z \otimes \boldsymbol{e}_x, \quad \boldsymbol{H}^{\mathrm{p}} = \gamma(x)\boldsymbol{e}_x \otimes \boldsymbol{e}_z \Rightarrow \boldsymbol{\varepsilon}^{\mathrm{e}} = (w'(x) - \gamma(x))\mathrm{sym}(\boldsymbol{e}_x \otimes \boldsymbol{e}_z). \tag{3.135}$$

Hence, the resolved shear stress is given by

$$\tau = \boldsymbol{\sigma} \cdot \boldsymbol{P}^{\mathrm{S}} = \sigma_{xz} = \sigma_{zx} \overset{(3.135)}{=} G(w'(x) - \gamma(x)). \tag{3.136}$$

Moreover, the linear momentum balance (3.119) implies

$$\partial_x \sigma_{zx} = 0 \quad \Rightarrow \quad \sigma_{xz} = \sigma_{zx} \overset{(3.132)_3}{=} \tau_{\mathrm{ext}} \overset{(3.136)}{=} \tau. \tag{3.137}$$

Obviously, the resolved shear stress τ is homogeneous and equal to the externally applied stress τ_{ext}. The micro moment equation (3.120) takes the form

$$\tau + \mathrm{div}\,(\boldsymbol{m}) \overset{(3.134)}{=} \tau_{\mathrm{ext}} + \frac{h^2}{12} E\gamma''(x) = 0. \tag{3.138}$$

The solution of this differential equation is given by

$$\gamma(x) = \frac{6\tau_{\mathrm{ext}}}{Eh^2}(D_{\mathrm{g}} x - x^2). \tag{3.139}$$

Here, the boundary conditions $(3.132)_2$ have been applied. A differential equation for $w(x)$ can be obtained by combining Eqns. (3.136), (3.137) and (3.139)

$$w'(x) = \gamma(x) + \frac{\tau_{\mathrm{ext}}}{G}. \tag{3.140}$$

Integration and application of the boundary condition $w(0) = 0$ gives

$$w(x) = \frac{\tau_{\mathrm{ext}}}{Eh^2}(3D_{\mathrm{g}} x^2 - 2x^3) + \frac{\tau_{\mathrm{ext}}}{G} x. \tag{3.141}$$

This is exactly the result (3.124), which has been obtained from the beam theory. Obviously, the shear influence is accounted for.

Finally, the density of geometrically necessary edge dislocations can be computed

$$\rho_{\vdash}(x) = -\gamma'(x) = \frac{6\tau_{\text{ext}}}{Eh^2}(2x - D_{\text{g}}). \tag{3.142}$$

Eq. (3.142) represents the dislocation pile-up structure. This result is not astonishing since the mathematical structure of the model at hand matches the structure of different gradient approaches like for example the theory of Gurtin et al. (2007). However, dislocation interactions have at no point explicitly been introduced in the current approach. Instead, they are assumed to be accounted for implicitly through the elastic micro-deformation of the layers.

Remark:

The grain boundaries of the model at hand are assumed perfectly micro-hard. This means that the rotations of the beam ends are constrained to vanish. This constraint leads to a non-vanishing bending moment at the grain boundaries, which increases with ongoing deformation. A more advanced model could allow for grain boundary yielding by introducing (at a certain point of the deformation process) additional slip bands such that those are matching at the grain boundaries. As a consequence, the plastic slip would no longer be constrained to vanish at the grain boundaries. An obvious criterion of the onset of new, matching slip bands may be the introduction of a critical (distributed) grain boundary micro moment. If the distributed micro moment at the grain boundary reaches this critical moment, the transition of the plastic deformation from one grain to another would then take place, i.e., the grain boundary would no longer behave micro hard. A familiar grain boundary model will be formulated in the subsequent chapter in the context of another gradient plasticity theory.

Chapter 4

Simplification of Gradient Plasticity based on a Scalar Plastic Field Variable

4.1 Introduction

The preceding chapters summarize various aspects of gradient crystal plasticity, for example, the kinematical description of GNDs and the consideration thereof in the constitutive framework. In this context, additional partial differential equations (PDEs) have been derived and interpreted as extra (micro force) balance equations (e.g. Eqns. (1.27), $(2.4)_2$ and (3.120)).

In the multislip case the number of additional scalar PDEs can be a multiple of the three scalar linear momentum balance equations. This implies a significant increase of the necessary computational effort compared to classical plasticity theories without internal length scale. In addition, the implementation of the discretized models is non-standard and non-trivial. For example, the determination of plastically active zones remains a demanding task within the context of gradient plasticity simulations. As a result, the application of associated models to three-dimensional problems remains an exception.

In addition, gradient plasticity models are currently far from being accurate. This is due to several uncertainties concerning, for example, boundary and interface conditions, the form of the potentials (see Chap. 2) and a lack of experimental data. Often, the predicted size effects are not only quantitatively wrong, but even qualitatively inaccurate. For example, the overall hardening of the model may be size-dependent.

However, many experiments rather suggest a size-dependent yield stress, instead. Therefore, the increase in computational cost and complexity usually contrasts with the gain of accuracy associated to gradient plasticity models.

The aforementioned computational effort may be reduced by decreasing the number of additional partial differential equations. This can be achieved by a modification of the kinematical framework. For example, one may not take into account the gradients of all plastic slips (or the curl of the plastic distorsion) but the gradient of only one scalar plastic field variable. This means that the kinematical information is reduced.

In the following, such a phenomenological, geometrically linear gradient plasticity framework including the gradient of a scalar plastic field variable is discussed. The theory can be considered as a gradient extended thermodynamic generalization of classical plasticity theories without internal length scale like the Von Mises theory (for an overview see, e.g., Simo and Hughes (1998)). It summarizes different aspects of gradient plasticity, like energetic or dissipative micro-stresses and interface models. These concepts are later applied to a gradient extension of single crystal plasticity. In order to understand the origin of the concepts as well as possible difficulties, they are first summarized in a more simple isotropic context.

Possible fields of application of the theory are for example micromechanical problems including size effects, interfaces like grain boundaries or the regularization of softening problems that otherwise suffer from localization. The model relies on one scalar plastic field variable ζ which is defined throughout the body under consideration. It is introduced in order to model the internal mechanical state of the material, in particular its resistance against plastic deformation as well as hardening or softening. The variable is supposed to have a kinematical character. In many applications, ζ has the interpretation of an equivalent plastic strain. It is kinematically coupled with the plastic strain through the flow rule. This is in line with many classical approaches (as summarized, e.g., by Simo and Hughes, 1998).

As already mentioned, the field ζ is intended to represent the internal mechanical state of the material. This means that a change in ζ is, for example, associated with a change of the energy state of the material or with dissipative processes. Therefore, generalized stresses are introduced extending power over $\dot{\zeta}$ and its gradient. These stresses are assumed to affect the plastic deformation. Therefore, they are accounted for in the yield condition and the flow rule. This implies the assumption that the aforementioned stresses (for example conjugate to ζ) control the onset of plastic deformation in addition to the Cauchy stress. This phenomenological approach allows, for example, to model hardening and, in the context of gradient theories, size effects. Taking into account not only ζ but also its gradient is considered as the main difference compared to the aforementioned classical theories.
Various motivations of gradient plasticity have been discussed in Chaps. 1, 2 and 3. Alternatively, the introduction of the gradient can be motivated by the assumption that the plastic response of a material point is in general nonlocal (Engelen et al., 2003). This means that the response does not exclusively depend on the internal state of that point but also on the surrounding particles. Simple nonlocal models can be approximated by gradient theories. For example, let the yield stress at a given point $\boldsymbol{x}$ be given by

$$\sigma_{\mathrm{Y}}(\boldsymbol{x}) = \sigma_0^{\mathrm{C}} + H\zeta(\boldsymbol{x}) + E \int_{\Delta V(\boldsymbol{x})} w(\|\boldsymbol{x}' - \boldsymbol{x}\|)\big(\zeta(\boldsymbol{x}) - \zeta(\boldsymbol{x}')\big)\,\mathrm{d}v'. \tag{4.1}$$

Here, σ_0^{C} is the initial yield stress and H and E denote the hardening and Young's modulus, respectively. The term $H\zeta(\boldsymbol{x})$ represents a standard local linear hardening model. The integral is a nonlocal extension. It introduces a plastic interaction of the considered particle at position $\boldsymbol{x}$ and its surrounding neighborhood $\Delta V(\boldsymbol{x})$. The spherical volume $\Delta V(\boldsymbol{x})$ is centered at $\boldsymbol{x}$ and w is a weighting function which decays to zero at $\partial\Delta V$. It is assumed normalized in the following sense

$$\int_{\Delta V(\boldsymbol{x})} w(\|\boldsymbol{x}' - \boldsymbol{x}\|)\,\mathrm{d}v' = 1. \tag{4.2}$$

The plastic particle interaction is established through the difference of the plastic variable ζ of the particle at $\boldsymbol{x}$ and any other particle at $\boldsymbol{x}'$ in the vicinity of $\boldsymbol{x}$. If ζ at $\boldsymbol{x}$ succeeds the value at $\boldsymbol{x}'$, the yield stress is (infinitesimally) increased. Otherwise, it is decreased. This means that a high plastic activity in the neighborhood of the particle at $\boldsymbol{x}$ 'stimulates' the plastic activity of the particle and vice versa. Therefore, plastically heterogeneous deformations are penalized by the integral term (this is the principal effect of the nonlocal extension). The yield stress can be approximated by replacing $\zeta(\boldsymbol{x}')$ by its Taylor series expansion around $\boldsymbol{x}$. If the series is truncated after the quadratic term, a short calculation yields

$$\sigma_{\mathrm{Y}}(\boldsymbol{x}) = \sigma_0^{\mathrm{C}} + H\zeta(\boldsymbol{x}) - El^2\Delta\zeta(\boldsymbol{x}). \tag{4.3}$$

Here, l is an internal length scale parameter, which has the order of magnitude of the averaging volume diameter. The influence of the Laplace term becomes generally important when the system size approaches the order of magnitude of l. Exceptions are given by localization phenomena. The relation (4.3) is an approximation of the nonlocal expression (4.1). Since Eq. (4.3) does not involve any integral quantities, but only an approximation of these based on the Laplacian, it is sometimes referred to as 'weakly nonlocal'.

Considering a generalized Von Mises theory and neglecting isotropic hardening, plastic yielding occurs if the yield condition is fulfilled

$$f = \|\boldsymbol{\sigma}'\| + El^2 \mathrm{div}\,(\nabla\zeta) - \sigma_0^{\mathrm{C}} = 0. \tag{4.4}$$

Formally, Eq. (4.4) is similar to the microforce balance (1.27) of Gurtin's single crystal gradient plasticity theory. In particular, for the quadratic approach $W_{\mathrm{g}} = El^2\sum_\alpha((\rho_{\vdash}^\alpha)^2 + (\rho_{\odot}^\alpha)^2)/2$, the balance equation (1.27) takes the following form

$$\tau_\alpha + El^2 \mathrm{div}\,(\boldsymbol{P}_\alpha^n \nabla\gamma_\alpha) - \tau_\alpha^{\mathrm{d}} = 0, \tag{4.5}$$

with the projectors $\boldsymbol{P}_\alpha^n = \boldsymbol{I} - \boldsymbol{n}_\alpha \otimes \boldsymbol{n}_\alpha$.

In addition, the term $El^2 \mathrm{div}\,(\nabla\zeta) = El^2\Delta\zeta$ is closely related to Groma's backstress (Eq. (2.57))

$$x \approx -\frac{c_0/b}{\rho} \partial^2_{x_1} \gamma. \tag{4.6}$$

The principal difference between both backstresses is given by the fact that the latter does not require the introduction of an additional material parameter. Instead, the internal length scale is given by $1/\sqrt{\rho}$, representing the average dislocation spacing.

In this sense, pragmatic and computationally manageable gradient plasticity models based on the scalar plastic field variable ζ can be motivated by general non-local theories, more complex gradient crystal plasticity frameworks as well as by the statistical dislocation theory of Groma et al. (2003).

In the following, phenomenological models which take the gradient of the scalar plastic field variable ζ into account, are revised in the thermodynamic context (neglecting temperature effects). The presentation comprises various aspects of phenomenological gradient plasticity theories

- The treatment by the principle of virtual power
- Boundary conditions and interfaces
- Energetic and dissipative micro-stresses
- Thermodynamic consistency
- Bulk and interface flow rules

These and other aspects have been discussed by various authors in the context of gradient plasticity. In order to formulate meaningful models for micromechanical systems, boundary and interface influences must not be neglected, if the interface-to-volume ratio reaches a critical value. Therefore, the possibility to formulate phenomenological plastic interface models is probably one of the most important features of gradient plasticity theories.

4.2 Power of the External and Internal Forces

The theory is based on the assumption of an additive decomposition of the strain into an elastic and a plastic part (Eq. (1.11))

$$\varepsilon = \operatorname{sym}(\nabla \boldsymbol{u}) = \varepsilon^{\mathrm{e}} + \varepsilon^{\mathrm{p}}, \tag{4.7}$$

where $\boldsymbol{u}$ denotes the displacement field. The body under consideration may consist of different phases

$$\mathcal{B} = \bigcup_i \mathcal{B}_i. \tag{4.8}$$

The phases are separated by interfaces, the union of which is denoted by Γ. The power of the internal forces is assumed to be given by

$$\mathcal{P}_{\text{int}} = \int_{\mathcal{B}} \left(\boldsymbol{\sigma} \cdot \dot{\varepsilon} + \xi \, \dot{\zeta} + \boldsymbol{\xi} \cdot \nabla \dot{\zeta} \right) \mathrm{d}v + \int_{\Gamma} \Xi_\Gamma \, \dot{\zeta} \, \mathrm{d}a. \tag{4.9}$$

Here, $\boldsymbol{\sigma}$ denotes the Cauchy stress conjugate to $\dot{\varepsilon}$ (i.e. $\boldsymbol{\sigma}$ extends power over $\dot{\varepsilon}$). Since changes of ζ and its gradient are assumed to be associated to internal thermodynamic processes, additional stresses ξ and $\boldsymbol{\xi}$ are assumed to extend power over $\dot{\zeta}$ and its gradient. The same holds for $\dot{\zeta}$ and Ξ_Γ at the interfaces. For simplicity, possible interface jumps of ζ are neglected. Just like the stress power $\boldsymbol{\sigma} \cdot \dot{\varepsilon}$, the power expressions $\xi \, \dot{\zeta}$ and $\boldsymbol{\xi} \cdot \nabla \dot{\zeta}$ as well as $\Xi_\Gamma \, \dot{\zeta}$ represent power densities. The associated energy is stored or dissipated inside the material. The power of the external forces is assumed to be given by

$$\mathcal{P}_{\text{ext}} = \int_{\partial \mathcal{B}} \boldsymbol{t} \cdot \dot{\boldsymbol{u}} \, \mathrm{d}a + \int_{\partial \mathcal{B}} \Xi \, \dot{\zeta} \, \mathrm{d}a. \tag{4.10}$$

Let the tractions $\boldsymbol{t}$ and microtractions Ξ be given by the fields $\bar{\boldsymbol{t}}$ and $\bar{\Xi}$ on the Neumann parts $\partial \mathcal{B}_t$ and $\partial \mathcal{B}_\Xi$ of the boundary, respectively. Body forces are neglected. On the Dirichlet parts $\partial \mathcal{B}_u$ and $\partial \mathcal{B}_\zeta$ of the boundary, $\boldsymbol{u}$ and ζ are assumed to be given. Due to energy conservation reasons it is assumed that

$$\mathcal{P}_{\text{int}} = \mathcal{P}_{\text{ext}}. \tag{4.11}$$

Note that at this point, there is no coupling between the conventional fields $\boldsymbol{\sigma}$ and ε and the purely phenomenological field ζ.

4.3 Principle of Virtual Power and Balance Equations

The relation (4.11) is assumed to hold for arbitrary processes. Such a process is considered in the following. It is assumed that the virtual rates $\delta\dot{\boldsymbol{u}}$ and $\delta\dot{\zeta}$ are introduced. Let these be compatible with the Dirichlet boundary conditions, i.e. $\delta\dot{\boldsymbol{u}} = \mathbf{0}$ on $\partial\mathcal{B}_u$ and $\delta\dot{\zeta} = 0$ on $\partial\mathcal{B}_\zeta$. In this case, one finds

$$\mathcal{P}_{\text{int}} + \delta\mathcal{P}_{\text{int}} = \mathcal{P}_{\text{ext}} + \delta\mathcal{P}_{\text{ext}}. \tag{4.12}$$

As Eq. (4.11) still holds, the following relation is obtained

$$\delta\mathcal{P}_{\text{int}} = \delta\mathcal{P}_{\text{ext}}. \tag{4.13}$$

This equation represents the principle of virtual power

$$\int_{\mathcal{B}} \left(\boldsymbol{\sigma} \cdot \delta\dot{\boldsymbol{\varepsilon}} + \xi\,\delta\dot{\zeta} + \boldsymbol{\xi} \cdot \nabla\delta\dot{\zeta}\right) \mathrm{d}v + \int_{\Gamma} \Xi_\Gamma\,\delta\dot{\zeta}\,\mathrm{d}a = \int_{\partial\mathcal{B}_t} \bar{\boldsymbol{t}} \cdot \delta\dot{\boldsymbol{u}}\,\mathrm{d}a + \int_{\partial\mathcal{B}_\Xi} \bar{\Xi}\,\delta\dot{\zeta}\,\mathrm{d}a. \tag{4.14}$$

Note that $\delta\dot{\boldsymbol{\varepsilon}}^{\mathrm{p}} = \mathbf{0}$, i.e., the variables $\boldsymbol{\varepsilon}^{\mathrm{p}}$ and ζ are chosen to be a priori independent.
Since the variations are arbitrary, one can deliberately choose $\delta\dot{\zeta} = 0$. The application of the chain rule and Gauss' theorem to Eq. (4.14) then yields

$$\int_{\mathcal{B}} \operatorname{div}(\boldsymbol{\sigma}) \cdot \delta\dot{\boldsymbol{u}}\,\mathrm{d}v = \int_{\partial\mathcal{B}_u} (\bar{\boldsymbol{t}} - \boldsymbol{\sigma}\boldsymbol{n}) \cdot \delta\dot{\boldsymbol{u}}\,\mathrm{d}a. \tag{4.15}$$

The arbitrariness of $\delta\dot{\boldsymbol{u}}$ implies

$$\operatorname{div}(\boldsymbol{\sigma}) = \mathbf{0} \quad \forall \boldsymbol{x} \in \mathcal{B}, \qquad \boldsymbol{\sigma}\boldsymbol{n} = \bar{\boldsymbol{t}} \quad \forall \boldsymbol{x} \in \partial\mathcal{B}_u. \tag{4.16}$$

These equations represent the linear momentum balance as well as Neumann-type boundary conditions.
An analogue procedure (choosing $\delta\dot{\boldsymbol{u}} = \mathbf{0}$ in Eq. (4.14)) yields a micro-force balance equation and associated Neumann boundary conditions. However, in this case the integration must be carried out for each

phase $\mathcal{B}_i$ separately due to the presence of the interface integral in Eq. (4.14). Consequently, interface conditions are obtained in addition

$$\xi = \operatorname{div}(\boldsymbol{\xi}) \;\; \forall \boldsymbol{x} \in \mathcal{B}_i, \qquad \Xi_\Gamma = [\![\boldsymbol{\xi}]\!] \cdot \boldsymbol{n} \;\; \forall \boldsymbol{x} \in \Gamma, \qquad \boldsymbol{\xi} \cdot \boldsymbol{n} = \bar{\Xi} \;\; \forall \boldsymbol{x} \in \partial\mathcal{B}_{\Xi}. \tag{4.17}$$

The jump of the normal component of the microstress $\boldsymbol{\xi}$ is defined by

$$[\![\boldsymbol{\xi}]\!] \cdot \boldsymbol{n} = (\boldsymbol{\xi}^+ - \boldsymbol{\xi}^-) \cdot \boldsymbol{n}, \tag{4.18}$$

where $\boldsymbol{n}$ points from '-' to '+'.

4.4 Constitutive Assumptions

4.4.1 Energetic Assumptions and Dissipation Inequality

The stored energy density is assumed to be a function of the elastic strain, the plastic variable ζ and $\nabla\zeta$

$$W = W(\boldsymbol{\varepsilon} - \boldsymbol{\varepsilon}^{\mathrm{p}}, \zeta, \nabla\zeta). \tag{4.19}$$

Moreover, the interface energy density is assumed to be a function of ζ

$$W_\Gamma = W_\Gamma(\zeta). \tag{4.20}$$

The constitutive model is required to be thermodynamically consistent. This means that the dissipation must be non-negative. The total dissipation is given by the difference of the power of the external forces (the energy which is introduced into the system) and the energy storage rate

$$D_{\mathrm{tot}} = \mathcal{P}_{\mathrm{ext}} - \int_{\mathcal{B}} \dot{W} \,\mathrm{d}v - \int_{\Gamma} \dot{W}_\Gamma \,\mathrm{d}a \geq 0. \tag{4.21}$$

From Eqns. (4.9), (4.11), (4.19) and (4.20) it follows that the total dissipation can be represented by

$$D_{\mathrm{tot}} = \sum_i \int_{\mathcal{B}_i} \Big((\boldsymbol{\sigma} - \partial_{\boldsymbol{\varepsilon}} W) \cdot \dot{\boldsymbol{\varepsilon}} + \partial_{\boldsymbol{\varepsilon}} W \cdot \dot{\boldsymbol{\varepsilon}}^{\mathrm{p}} + (\xi - \partial_\zeta W)\dot{\zeta} + (\boldsymbol{\xi} - \partial_{\nabla\zeta} W) \cdot \nabla\dot{\zeta} \Big) \,\mathrm{d}v$$
$$+ \int_{\Gamma} (\Xi_\Gamma - \partial_\zeta W_\Gamma)\dot{\zeta} \,\mathrm{d}a. \tag{4.22}$$

The integrand of the first integral represents the local (bulk) dissipation

$$\mathcal{D} = (\boldsymbol{\sigma} - \partial_{\varepsilon} W) \cdot \dot{\boldsymbol{\varepsilon}} + \partial_{\varepsilon} W \cdot \dot{\boldsymbol{\varepsilon}}^{\mathrm{p}} + (\xi - \partial_{\zeta} W)\dot{\zeta} + (\boldsymbol{\xi} - \partial_{\nabla\zeta} W) \cdot \nabla\dot{\zeta} \geq 0. \quad (4.23)$$

The dissipation is required to be positive or zero. Assuming that, for virtually and instantaneously frozen states $\boldsymbol{\varepsilon}^{\mathrm{p}}$, ζ and $\nabla\zeta$, the stress power $\boldsymbol{\sigma} \cdot \dot{\boldsymbol{\varepsilon}}$ is fully stored energetically, one finds for arbitrary strain rates $\dot{\boldsymbol{\varepsilon}}$

$$\mathcal{D}(\dot{\boldsymbol{\varepsilon}}^{\mathrm{p}} = \mathbf{0},\ \dot{\zeta} = 0,\ \nabla\dot{\zeta} = \mathbf{0}) = (\boldsymbol{\sigma} - \partial_{\varepsilon} W) \cdot \dot{\boldsymbol{\varepsilon}} = 0. \quad (4.24)$$

This statement implies the following constitutive relation for the Cauchy stress

$$\boldsymbol{\sigma} = \partial_{\varepsilon} W. \quad (4.25)$$

For better readability, the following energetic and dissipative stresses are defined

$$\xi^{\mathrm{e}} = \partial_{\zeta} W,\ \ \xi^{\mathrm{d}} = \xi - \xi^{\mathrm{e}} \overset{(4.17)}{=} \mathrm{div}\,(\boldsymbol{\xi}) - \partial_{\zeta} W,\ \ \boldsymbol{\xi}^{\mathrm{e}} = \partial_{\nabla\zeta} W,\ \ \boldsymbol{\xi}^{\mathrm{d}} = \boldsymbol{\xi} - \boldsymbol{\xi}^{\mathrm{e}}. \quad (4.26)$$

The stress powers of the energetic stresses and their conjugate kinematic rates, for example $\xi^{\mathrm{e}}\dot{\zeta} = \partial_{\zeta} W \dot{\zeta}$, lead to energy storage (or release) in the material. In contrast, the sum of the stress powers of the dissipative stresses and the associated rates gives the dissipation

$$\mathcal{D} = \boldsymbol{\sigma} \cdot \dot{\boldsymbol{\varepsilon}}^{\mathrm{p}} + \xi^{\mathrm{d}}\,\dot{\zeta} + \boldsymbol{\xi}^{\mathrm{d}} \cdot \nabla\dot{\zeta} \ \geq\ 0. \quad (4.27)$$

Equation (4.27) is the reduced dissipation inequality of the bulk. At the interface, an analogous procedure yields the dissipation per unit area

$$\mathcal{D}_{\Gamma} = (\Xi_{\Gamma} - \Xi_{\Gamma}^{\mathrm{e}})\,\dot{\zeta} \geq 0 \quad \text{with}\ \Xi_{\Gamma}^{\mathrm{e}} = \partial_{\zeta} W_{\Gamma}. \quad (4.28)$$

For better readability, the dissipative interface microtraction Ξ^{d} is introduced

$$\Xi_{\Gamma}^{\mathrm{d}} = \Xi_{\Gamma} - \Xi_{\Gamma}^{\mathrm{e}} \ \Rightarrow\ \mathcal{D}_{\Gamma} = \Xi^{\mathrm{d}}\,\dot{\zeta}. \quad (4.29)$$

4.4.2 Rate-Independent Setting

The reduced dissipation inequality (4.27) states that the dissipative stresses can be interpreted as the driving forces of dissipative processes like the plastic deformation. Therefore, the plastic deformation is assumed to be triggered when the dissipative stresses reach a critical value. This relation is expressed by the yield criterion which is assumed to be given by

$$f = f(\boldsymbol{\sigma}, \xi^{\mathrm{d}}, \boldsymbol{\xi}^{\mathrm{d}}) \leq 0. \tag{4.30}$$

In the case of plastic yielding $f = 0$ holds.

Remark: In many applications the stored energy is assumed to have an additive form

$$W = W_{\mathrm{e}}(\boldsymbol{\varepsilon} - \boldsymbol{\varepsilon}^{\mathrm{p}}) + W_{\mathrm{h}}(\zeta) + W_{\mathrm{g}}(\nabla\zeta). \tag{4.31}$$

Interestingly, in this case, no equations exist in the theory which couple the fields $\{\boldsymbol{u}, \boldsymbol{\varepsilon}, \boldsymbol{\varepsilon}^{\mathrm{p}}\}$ with the field ζ until the yield condition is introduced.

The dissipation inequality (4.27) can be exploited to identify suitable constitutive equations involving $\dot{\boldsymbol{\varepsilon}}^{\mathrm{p}}$, $\dot{\zeta}$ and $\boldsymbol{\xi}^{\mathrm{d}}$. These constitutive models must satisfy Eq. (4.27). This can be realized by maximizing the dissipation. Therefore, the following Lagrangian is considered

$$\mathcal{F}_{\mathrm{L}}(\boldsymbol{\sigma}, \xi^{\mathrm{d}}, \boldsymbol{\xi}^{\mathrm{d}}) = \boldsymbol{\sigma} \cdot \dot{\boldsymbol{\varepsilon}}^{\mathrm{p}} + \xi^{\mathrm{d}} \dot{\zeta} + \boldsymbol{\xi}^{\mathrm{d}} \cdot \nabla\dot{\zeta} - \mathring{\lambda} f(\boldsymbol{\sigma}, \xi^{\mathrm{d}}, \boldsymbol{\xi}^{\mathrm{d}}). \tag{4.32}$$

Accordingly, for given non-vanishing plastic rates $\{\dot{\boldsymbol{\varepsilon}}^{\mathrm{p}}, \dot{\zeta}, \nabla\dot{\zeta}\}$ one assumes that the associated stress state $\{\boldsymbol{\sigma}, \xi^{\mathrm{d}}, \boldsymbol{\xi}^{\mathrm{d}}\}$ maximizes the dissipation. The yield condition $f = 0$ takes the role of a constraint on the stress state and is accounted for by means of the Lagrange multiplier $\mathring{\lambda}$. The symbol $(\mathring{\bullet})$ expresses the rate-like character of the multiplier. From the dissipation maximization requirement, it follows that

$$\partial_{\sigma}\mathcal{F}_{\mathrm{L}} = \mathbf{0} \;\Rightarrow\; \dot{\boldsymbol{\varepsilon}}^{\mathrm{p}} = \mathring{\lambda}\partial_{\sigma} f, \tag{4.33}$$

$$\partial_{\xi^{\mathrm{d}}}\mathcal{F}_{\mathrm{L}} = 0 \;\Rightarrow\; \dot{\zeta} = \mathring{\lambda}\partial_{\xi^{\mathrm{d}}} f, \tag{4.34}$$

$$\partial_{\boldsymbol{\xi}^{\mathrm{d}}}\mathcal{F}_{\mathrm{L}} = \mathbf{0} \;\Rightarrow\; \nabla\dot{\zeta} = \mathring{\lambda}\partial_{\boldsymbol{\xi}^{\mathrm{d}}} f. \tag{4.35}$$

This set of equations is associated to the unknowns $\dot{\varepsilon}^{\mathrm{p}}$, $\dot{\zeta}$ and $\boldsymbol{\xi}^{\mathrm{d}}$. The first and second equations can be interpreted as flow rules for $\dot{\varepsilon}^{\mathrm{p}}$ and $\dot{\zeta}$ which thereby become kinematically coupled. Note that the principle of maximum dissipation leaves no freedom in the choice of these flow rules. As a consequence, the kinematical coupling of ζ with the plastic strain is completely determined through the yield criterion, if the principle of maximum dissipation is applied.
Let the yield criterion have the form

$$f(\boldsymbol{\sigma}, \xi^{\mathrm{d}}, \boldsymbol{\xi}^{\mathrm{d}}) = \tilde{f}(\boldsymbol{\sigma}, \xi^{\mathrm{d}}, \boldsymbol{\xi}^{\mathrm{d}}) - \tilde{c}, \tag{4.36}$$

where $\tilde{f}$ is homogeneous of degree one with respect to the stresses and $\tilde{c} \geq 0$ is a constant. Equation (4.36) implies the following representation of the dissipation (Miehe, 2011)

$$\mathcal{D} = (\boldsymbol{\sigma} \cdot \partial_{\sigma}\tilde{f} + \xi^{\mathrm{d}}\,\partial_{\xi^{\mathrm{d}}}\tilde{f} + \boldsymbol{\xi}^{\mathrm{d}} \cdot \partial_{\boldsymbol{\xi}^{\mathrm{d}}}\tilde{f})\mathring{\lambda} = \tilde{c}\,\mathring{\lambda} \geq 0. \tag{4.37}$$

The Kuhn-Tucker conditions summarize the results and take the classical form

$$\mathring{\lambda} \geq 0, \quad f \leq 0, \quad \mathring{\lambda} f = 0. \tag{4.38}$$

Considering the interface dissipation (4.29), a canonical form of the interface flow rule is given by

$$f_{\Gamma} = \Xi_{\Gamma}^{\mathrm{d}} - \Xi_0^{\mathrm{C}} \overset{(4.29)}{=} \Xi_{\Gamma} - (\Xi_0^{\mathrm{C}} + \Xi_{\Gamma}^{\mathrm{e}}), \tag{4.39}$$

with the initial yield strength $\Xi_0^{\mathrm{C}} \geq 0$. The microtraction $\Xi_{\Gamma} = [\![\boldsymbol{\xi}]\!] \cdot \boldsymbol{n}$ (see Eq. (4.17)) can be interpreted as the interface loading. The energetic microtraction $\Xi_{\Gamma}^{\mathrm{e}} = \partial_{\zeta} W_{\Gamma}$ (Eq. (4.29)) determines the hardening behavior of the interface. The interface has a micro-hard behavior for $f_{\Gamma} < 0$. This

means, that in this case, the interface condition $\dot{\zeta} = 0$ has a Dirichlet character. In contrast, interface yielding implies $f_\Gamma = 0$, at least in the rate-independent setting.
Given $f_\Gamma = 0$, the loading condition is given by $\dot{f}_\Gamma\big|_\zeta > 0$. If the loading condition is fulfilled, the interface conditions, given by $[\![\boldsymbol{\xi}]\!] \cdot \boldsymbol{n} = \Xi_0^{\mathrm{C}} + \Xi_\Gamma^{\mathrm{e}}$, exhibit a mixed Neumann-Robin character. Finally, the dissipation inequality $\mathcal{D}_\Gamma = \Xi_\Gamma^{\mathrm{d}} \dot{\zeta} = \Xi_0^{\mathrm{C}} \dot{\zeta} \geq 0$ implies $\dot{\zeta} \geq 0$ on Γ. As a summary, interface conditions which are in line with the Kuhn-Tucker conditions (4.38) are obtained

$$\dot{\zeta} \geq 0, \quad f_\Gamma \leq 0, \quad \dot{\zeta} f_\Gamma = 0. \tag{4.40}$$

Example: Let the yield criterion be given by (compare, e.g., Fleck and Willis, 2009a)

$$f(\boldsymbol{\sigma}, \xi^{\mathrm{d}}, \boldsymbol{\xi}^{\mathrm{d}}) = \sqrt{\|\boldsymbol{\sigma}'\|^2 + (1/l^{\mathrm{d}})^2 \|\boldsymbol{\xi}^{\mathrm{d}}\|^2} + \sqrt{\frac{2}{3}}(\xi^{\mathrm{d}} - \sigma_0^{\mathrm{C}}), \tag{4.41}$$

where σ_0^{C} is the initial yield strength and l^{d} represents an internal length scale parameter. Eqns. (4.33) to (4.35) take the form

$$\dot{\boldsymbol{\varepsilon}}^{\mathrm{p}} = \overset{\circ}{\lambda} \frac{\boldsymbol{\sigma}'}{\sqrt{\|\boldsymbol{\sigma}'\|^2 + (1/l^{\mathrm{d}})^2 \|\boldsymbol{\xi}^{\mathrm{d}}\|^2}}, \tag{4.42}$$

$$\dot{\zeta} = \sqrt{\frac{2}{3}} \overset{\circ}{\lambda}, \tag{4.43}$$

$$l^{\mathrm{d}} \nabla \dot{\zeta} = \overset{\circ}{\lambda} \frac{\boldsymbol{\xi}^{\mathrm{d}}}{\sqrt{(l^{\mathrm{d}})^2 \|\boldsymbol{\sigma}'\|^2 + \|\boldsymbol{\xi}^{\mathrm{d}}\|^2}}. \tag{4.44}$$

Note that for large mechanical systems without localization effects, one can generally assume that $\|\dot{\boldsymbol{\varepsilon}}^{\mathrm{p}}\| \gg \|l^{\mathrm{d}} \nabla \dot{\zeta}\|$ which implies $\|\boldsymbol{\sigma}'\| \gg \|\boldsymbol{\xi}^{\mathrm{d}}\|/l^{\mathrm{d}}$. Hence, the dissipative gradient effect incorporated by $\boldsymbol{\xi}^{\mathrm{d}}$ is negligible if the system length scale is significantly larger than l^{d}.
With Eq. (4.26) the yield condition can be represented by

$$f(\boldsymbol{\sigma}, \xi^{\mathrm{d}}, \boldsymbol{\xi}^{\mathrm{d}}) = \left(\sqrt{\|\boldsymbol{\sigma}'\|^2 + (1/l^{\mathrm{d}})^2 \|\boldsymbol{\xi}^{\mathrm{d}}\|^2} + \sqrt{\frac{2}{3}} \operatorname{div}(\boldsymbol{\xi}^{\mathrm{e}}) \right) - \sqrt{\frac{2}{3}}(\sigma_0^{\mathrm{C}} + \xi^{\mathrm{e}}). \tag{4.45}$$

In order to get more specific, the following energy is assumed

$$W = \frac{1}{2}\boldsymbol{\varepsilon}^{\rm e} \cdot \mathbb{C}[\boldsymbol{\varepsilon}^{\rm e}] + \frac{1}{2}H\zeta^2 + \frac{1}{2}(l^{\rm e})^2 E\nabla\zeta \cdot \nabla\zeta \tag{4.46}$$

with the linear hardening modulus H and the energetic length scale parameter $l^{\rm e}$. Neglecting dissipative gradient effects, the yield condition (4.45) becomes

$$f = \|\boldsymbol{\sigma}'\| - \sqrt{\frac{2}{3}}\big(\sigma_0^{\rm C} + H\zeta - E(l^{\rm e})^2\Delta\zeta\big). \tag{4.47}$$

This yield criterion is equivalent to the yield stress (4.3) which has been obtained from the approximation of a nonlocal theory.
If gradient influences are neglected, the yield condition (4.45) equals the classical Von Mises yield condition with isotropic hardening stress $\xi^{\rm e} = \partial_\zeta W$.

4.4.3 Rate-Dependent Setting

In the rate-depend setting, stress states $\{\boldsymbol{\sigma}, \xi^{\rm d}, \boldsymbol{\xi}^{\rm d}\}$ in the region $f(\boldsymbol{\sigma}, \xi^{\rm d}, \boldsymbol{\xi}^{\rm d}) > 0$ are generally possible. They are called overstress states and determine the plastic deformation rate. It is assumed that Eqns. (4.33) to (4.35) remain formally identical

$$\dot{\boldsymbol{\varepsilon}}^{\rm p} = \mathring{\lambda}\partial_{\boldsymbol{\sigma}} f, \tag{4.48}$$
$$\dot{\zeta} = \mathring{\lambda}\partial_{\xi^{\rm d}} f, \tag{4.49}$$
$$\nabla\dot{\zeta} = \mathring{\lambda}\partial_{\boldsymbol{\xi}^{\rm d}} f. \tag{4.50}$$

However, $\mathring{\lambda}$ is no longer a Lagrange multiplier but a monotonous function of the overstress state

$$\mathring{\lambda} = \mathring{\lambda}\big(\langle f\rangle\big) \geq 0, \quad \text{with } \mathring{\lambda}(0) = 0, \tag{4.51}$$

where $\langle \bullet \rangle = \big(|(\bullet)| + (\bullet)\big)/2$ denote the McAuley brackets. Many applications are based on a power-type creep law

$$\mathring{\lambda} = \dot{\lambda}_0 \left\langle \frac{f}{\sigma^{\rm D}} \right\rangle^p. \tag{4.52}$$

The reference strain rate $\dot{\lambda}_0$, the drag stress σ^{D} and the strain rate sensitivity p are material parameters. Note that $\dot{\lambda}_0$ and σ^{D} are not independent (i.e. a multiplication of σ^{D} by a factor a has the same effect as a division of $\dot{\lambda}_0$ by a^p).
The dissipation (4.27) is given by (compare Eq. (4.36))

$$\mathcal{D} = (\boldsymbol{\sigma} \cdot \partial_{\sigma} \tilde{f} + \xi^{\mathrm{d}} \, \partial_{\xi^{\mathrm{d}}} \tilde{f} + \boldsymbol{\xi}^{\mathrm{d}} \cdot \partial_{\boldsymbol{\xi}^{\mathrm{d}}} \tilde{f}) \mathring{\lambda} = (f + \sigma_0^{\mathrm{C}}) \, \mathring{\lambda} \geq 0. \tag{4.53}$$

Here, the value of f represents the overstress, which has a viscous character.
The rate-depend interface theory follows the same line as the bulk theory. It is based on the following rate equation

$$\dot{\zeta} = \mathring{\lambda}_{\Gamma}(\langle f_{\Gamma} \rangle) \overset{(4.39)}{=} \mathring{\lambda}_{\Gamma}(\langle [\![\boldsymbol{\xi}]\!] \cdot \boldsymbol{n} - (\Xi_0^{\mathrm{C}} + \Xi_{\Gamma}^{\mathrm{e}}) \rangle) \text{ on } \Gamma, \tag{4.54}$$

where the function $\mathring{\lambda}_{\Gamma}(\bullet)$ has the same properties as $\mathring{\lambda}(\bullet)$. Eq. (4.54) is again based on the overstress f_{Γ} at the interface, which also has a viscous character.
Note that a rate-dependent bulk model does not preclude rate-independent interface equations and vice versa.

4.5 Conceptual Difficulties in Plasticity including the Gradient of one Scalar Plastic Field Variable

4.5.1 One-Dimensional Example

A conceptual difficulty in plasticity including the gradient of one scalar plastic field variable ζ arises, e.g., in the following situation

- the generalized dissipative stress $\boldsymbol{\xi}^{\mathrm{d}}$ is neglected
- the yield criterion $f = 0$ is satisfied at a given point
- at the same time, the Cauchy stress is zero

This situation occurs, for example, when the term $\mathrm{div}\,(\boldsymbol{\xi}^{\mathrm{e}})$ in Eq. (4.45) is sufficiently large. In this case, plastic deformation is triggered, but the flow rule (4.42) cannot provide the direction of $\dot{\varepsilon}^{\mathrm{p}}$. In order to further

investigate this problem, a one-dimensional viscoplastic example is considered in the following.
The stored energy function is assumed to be given by

$$W = \frac{1}{2}E(\varepsilon - \varepsilon^{\mathrm{p}})^2 + \frac{1}{2}K_{\mathrm{G}}(\zeta')^2 \text{ with } \zeta' = \zeta'(x),\ K_{\mathrm{G}} = E(l^{\mathrm{e}})^2. \qquad (4.55)$$

Neglecting dissipative microstresses, the yield criterion reads (compare Eq. (4.45))

$$f = |\sigma| + K_{\mathrm{G}}\zeta'' - \sigma_0^{\mathrm{C}}. \qquad (4.56)$$

The flow rule (4.42) in combination with the creep law approach (4.52) takes the form

$$\dot{\varepsilon}^{\mathrm{p}} = \dot{\varepsilon}_0 \operatorname{sgn}(\sigma) \left\langle \frac{f}{\sigma^{\mathrm{D}}} \right\rangle^p = \dot{\varepsilon}_0 \operatorname{sgn}(\sigma) \left\langle \frac{|\sigma| + K_{\mathrm{G}}\zeta'' - \sigma_0^{\mathrm{C}}}{\sigma^{\mathrm{D}}} \right\rangle^p \qquad (4.57)$$

with the reference strain rate $\dot{\varepsilon}_0$. The field ζ satisfies the equation $\dot{\zeta} = |\dot{\varepsilon}^{\mathrm{p}}|$ (compare Eq. (4.49)). For simplicity, the initial yield stress is set to zero and the rate sensitivity is chosen to be linear, i.e. $\sigma_0^{\mathrm{C}} = 0$ and $p = 1$. Hence, the model is reduced to a viscoelastic gradient model. The viscoelastic setting is sufficient to illustrate the aforementioned conceptual difficulty, which is discussed by means of the following initial boundary value problem.

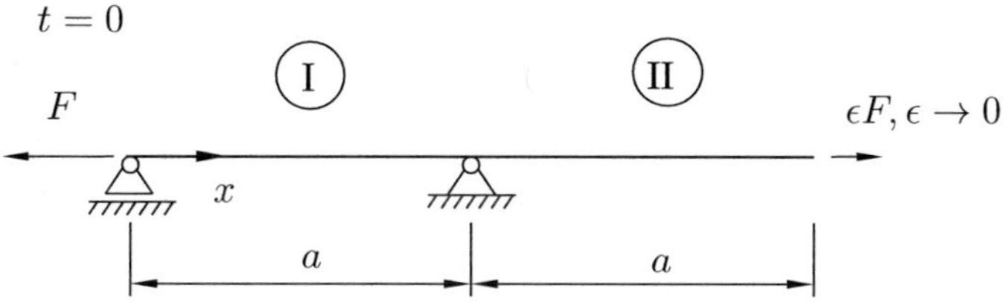

Figure 4.1: Initial configuration of the bar. The direction of the force ϵF is determined by the sign of ϵ.

A bar (see Fig. 4.1) is loaded by a tensile force $F > 0$ and an additional small perturbation force ϵF. The direction of the perturbation is determined by $\operatorname{sgn}(\epsilon)$. Here, the limit process $\epsilon \to 0$ is of special interest. The stress in the two regions is given by

$$\text{Region } I(0 \leq x \leq a): \quad \sigma^I = \frac{F}{A}, \tag{4.58}$$

$$\text{Region } II(a < x \leq 2a): \quad \sigma^{II} = \frac{\epsilon F}{A}. \tag{4.59}$$

Here, A denotes the cross sectional area of the bar.
Let the initial conditions be given by

$$\varepsilon^{\mathrm{p}I}(x, t = 0) = \zeta^I(x, t = 0) = \frac{Fa^2}{4AK_{\mathrm{G}}}\left(2 - \left(\frac{x}{a}\right)^2\right), \tag{4.60}$$

$$\varepsilon^{\mathrm{p}II}(x, t = 0) = \zeta^{II}(x, t = 0) = \frac{Fa^2}{4AK_{\mathrm{G}}}\left(\frac{x - 2a}{a}\right)^2. \tag{4.61}$$

These initial conditions allow for a simple solution of the problem.
It is assumed that the loading F is small and consequently the pre-deformation is small, too. In the one-dimensional setting, the field equations (4.16) and (4.17) take the form

$$\sigma' = 0, \quad \xi = K_{\mathrm{G}}\zeta''. \tag{4.62}$$

Let the boundary conditions be given by

$$(\zeta^I)'(x = 0) = (\zeta^{II})'(x = 2a) = 0. \tag{4.63}$$

This means that microfree conditions are assumed at the boundaries. The support at $x = a$ has no direct effect on the field ζ since it can deliberately be replaced by the associated support reaction. For $\epsilon \to 0$, the solution of the problem is given by

$$\varepsilon^{\mathrm{p}I} = \zeta^I = \frac{F\dot{\varepsilon}_0}{2\sigma^{\mathrm{D}}A}t + \varepsilon^{\mathrm{p}I}(x, 0) \tag{4.64}$$

$$\varepsilon^{\mathrm{p}II} = \operatorname{sgn}(\epsilon)\frac{F\dot{\varepsilon}_0}{2\sigma^{\mathrm{D}}A}t + \varepsilon^{\mathrm{p}II}(x, 0) \tag{4.65}$$

$$\zeta^{II} = \frac{F\dot{\varepsilon}_0}{2\sigma^{\mathrm{D}}A}t + \varepsilon^{\mathrm{p}II}(x, 0). \tag{4.66}$$

The displacement field can easily be obtained by integration, but is not of interest here. For sufficiently large times t, the strain is approximately given by ε^{p}. The deformation in region I is characterized by tensile strains. The sign of the strain in region II is determined by the sign of the perturbation ϵF. This means that the perturbation ϵF has a significant influence on the solution. It determines if the right part of the bar is stretched or compressed. This is illustrated in Fig. 4.2.

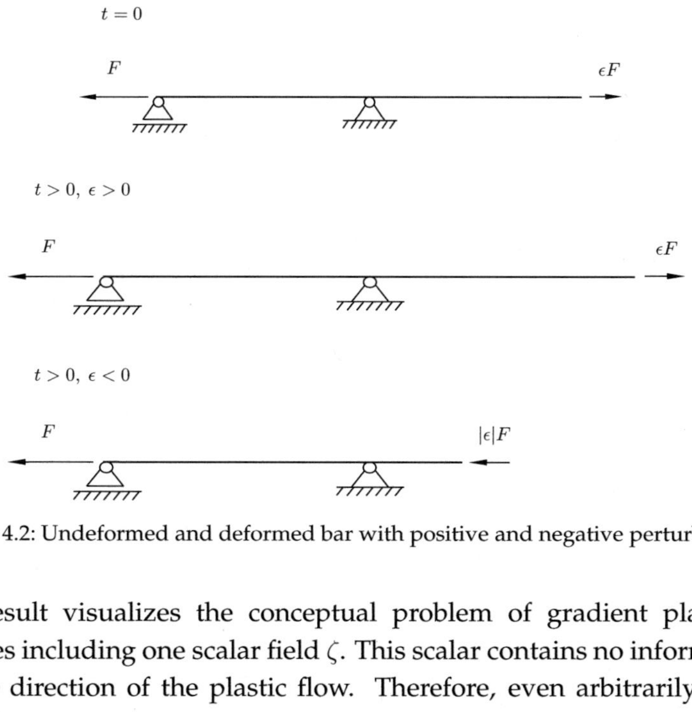

Figure 4.2: Undeformed and deformed bar with positive and negative perturbation.

The result visualizes the conceptual problem of gradient plasticity theories including one scalar field ζ. This scalar contains no information on the direction of the plastic flow. Therefore, even arbitrarily small perturbations can determine the direction of the plastic deformation. In this sense, the solution is not stable with respect to the boundary conditions.

4.5.2 A Stable One-Dimensional Formulation

A stable one-dimensional formulation can be obtained by accounting for tensile and compressive plastic strain increments separately. Consider for example the following flow rule for a material without gradient effects (formulation A)

$$\dot{\varepsilon}^{\rm p} = {\rm sgn}\,(\sigma)\,\dot{\varepsilon}_0 \left\langle \frac{|\sigma| - \sigma_0^{\rm C}}{\sigma^{\rm D}} \right\rangle . \tag{4.67}$$

An equivalent formulation can be obtained as follows. Let the flow rules for tensile and compressive plastic strain increments be given by (formulation B)

$$\dot{\varepsilon}_{\rm t}^{\rm p} = \dot{\varepsilon}_0 \left\langle \frac{\sigma - \sigma_0^{\rm C}}{\sigma^{\rm D}} \right\rangle, \quad \dot{\varepsilon}_{\rm c}^{\rm p} = \dot{\varepsilon}_0 \left\langle \frac{-\sigma - \sigma_0^{\rm C}}{\sigma^{\rm D}} \right\rangle . \tag{4.68}$$

These flow rules implicitly introduce the following yield criteria for tensile and compressive plastic flow

$$f_{\rm t} = \sigma - \sigma_0^{\rm C}, \quad f_{\rm c} = -\sigma - \sigma_0^{\rm C}. \tag{4.69}$$

In this formulation, the flow rule for $\dot{\varepsilon}^{\rm p}$ is given by the sum of the tensile and compressive strain rates

$$\dot{\varepsilon}^{\rm p} = \dot{\varepsilon}_{\rm t}^{\rm p} - \dot{\varepsilon}_{\rm c}^{\rm p} \Rightarrow \varepsilon^{\rm p} = \varepsilon_{\rm t}^{\rm p} - \varepsilon_{\rm c}^{\rm p}. \tag{4.70}$$

It can be verified that this formulation is consistent with Eq. (4.67). This means, that both formulations, A and B, are equivalent if no gradient extension is introduced. However, this does not hold for the gradient extended versions. The gradient extended version of formulation A has been discussed in the last section. It has been shown to possibly yield instable solutions. The gradient extended version of formulation B reads (compare Eq. (4.52))

$$\dot{\varepsilon}_{\rm t}^{\rm p} = \dot{\varepsilon}_0 \left\langle \frac{\sigma + K_{\rm G}\zeta'' - \sigma_0^{\rm C}}{\sigma^{\rm D}} \right\rangle, \quad \dot{\varepsilon}_{\rm c}^{\rm p} = \dot{\varepsilon}_0 \left\langle \frac{-\sigma + K_{\rm G}\zeta'' - \sigma_0^{\rm C}}{\sigma^{\rm D}} \right\rangle . \tag{4.71}$$

The associated yield criteria for tensile and compressive plastic deformations are in line with Eq. (4.45)

$$f_{\rm t} = \sigma + K_{\rm G}\zeta'' - \sigma_0^{\rm C}, \quad f_{\rm c} = -\sigma + K_{\rm G}\zeta'' - \sigma_0^{\rm C}. \tag{4.72}$$

The flow rule for ζ consequently reads (comp. Eq. (4.49))

$$\dot{\zeta} = \dot{\varepsilon}_{\mathrm{t}}^{\mathrm{p}} + \dot{\varepsilon}_{\mathrm{c}}^{\mathrm{p}} = \dot{\varepsilon}_0 \left\langle \frac{\sigma + K_{\mathrm{G}}\zeta'' - \sigma_0^{\mathrm{C}}}{\sigma^{\mathrm{D}}} \right\rangle + \dot{\varepsilon}_0 \left\langle \frac{-\sigma + K_{\mathrm{G}}\zeta'' - \sigma_0^{\mathrm{C}}}{\sigma^{\mathrm{D}}} \right\rangle . \tag{4.73}$$

In contrast to formulation A, the flow rules of formulation B do not require the evaluation of $\mathrm{sgn}\,(\sigma)$. Remember that the factor $\mathrm{sgn}\,(\sigma)$ has been at the origin of the instable behavior of formulation A. This has been illustrated by the example in the last section. If the same problem is solved with formulation B, it can easily be verified that the deformation of the right part of the bar vanishes for $\epsilon \to 0$, since

$$\dot{\varepsilon}^{\mathrm{p}II} = \dot{\varepsilon}_{\mathrm{t}}^{\mathrm{p}II} - \dot{\varepsilon}_{\mathrm{c}}^{\mathrm{p}II} \overset{(4.71)}{=} \dot{\varepsilon}_0 \left\langle \frac{\epsilon F/A + K_{\mathrm{G}}\zeta''}{\sigma^{\mathrm{D}}} \right\rangle - \dot{\varepsilon}_0 \left\langle \frac{-\epsilon F/A + K_{\mathrm{G}}\zeta''}{\sigma^{\mathrm{D}}} \right\rangle \underset{0}{\overset{\epsilon}{\to}} 0. \tag{4.74}$$

Therefore, the instable behavior of formulation A does not appear when formulation B is applied. The instability does not emerge using formulation B since the gradient term $K_{\mathrm{G}}\zeta''$ equally enters the flow rules for tensile and compressive plastic strains. This is conceptually in line with the fact that the scalar variable ζ contains no information on the direction of the plastic strain rate, but only on its amount.

Discussion. The instable character of the gradient model vanishes if formulation B is applied. However, the introduction of separate flow rules for tensile and compressive plastic strains seems artificial. Moreover, it can be shown that the dissipation of formulation B and the rate $\dot{\zeta}$ can take finite values in the case $\dot{\varepsilon}^{\mathrm{p}} = 0$. These unphysical properties of the model occur if the term $K_{\mathrm{G}}\zeta''$ becomes too large.
These disadvantages are interpreted as unavoidable consequences of the simplicity of the model, incorporated by the gradient of only one single scalar plastic field variable ζ. It can be argued that more complex models based on the full dislocation density tensor $\boldsymbol{\alpha}$ do not suffer from the aforementioned drawbacks (see also the work of Poh et al., 2011). However, in many applications the aforementioned drawbacks of the simplified theory play a minor role. As already mentioned, the problems occur if the term $K_{\mathrm{G}}\zeta''$ becomes large. However, very often this term is

negative, especially in the region of micro-hard boundaries or interfaces. In these cases, the aforementioned disadvantages are often negligible.

4.6 Rate-Dependent Gradient Crystal Plasticity

The theory introduced in Sects. 4.3 and 4.4 can be generalized to a crystal plasticity setting by the introduction of yield criteria for the different slip systems. These yield criteria are based on the resolved shear stresses

$$\tau_\alpha = \boldsymbol{\sigma} \cdot \boldsymbol{P}^{\mathrm{S}}_\alpha, \quad \text{with } \boldsymbol{P}^{\mathrm{S}}_\alpha = \mathrm{sym}(\boldsymbol{d}_\alpha \otimes \boldsymbol{n}_\alpha). \tag{4.75}$$

Here, $\boldsymbol{d}_\alpha$ and $\boldsymbol{n}_\alpha$ denote the slip directions and slip plane normals, respectively.
For simplicity, dissipative microstresses are neglected

$$\boldsymbol{\xi}^{\mathrm{d}} = \boldsymbol{0} \ \Rightarrow \boldsymbol{\xi} = \boldsymbol{\xi}^{\mathrm{e}}. \tag{4.76}$$

The yield criteria are assumed to be given by

$$f_\alpha = \tau_\alpha + \xi^{\mathrm{d}} - \tau_0^{\mathrm{C}}, \tag{4.77}$$

where τ_0^{C} denotes the initial yield stress. This representation is in line with Eq. (4.36). In order to avoid instable solutions, two yield criteria are introduced for each crystallographic slip system. These yield criteria account for the positive and negative slip directions. Here, the following convention is used for opposed slip directions $\boldsymbol{d}_\alpha$ and $\boldsymbol{d}'_\alpha$

$$\boldsymbol{d}_\alpha = -\boldsymbol{d}'_\alpha, \quad \boldsymbol{d}'_\alpha = \boldsymbol{d}_{\alpha+N}, \quad \alpha = 1, ..., N \tag{4.78}$$

with N being the number of slip systems. The slip plane normals of opposed slip directions are equal $\boldsymbol{n}_\alpha = \boldsymbol{n}'_\alpha = \boldsymbol{n}_{\alpha+N}$. For example, face centered cubic (FCC) crystals, which have 12 slip systems, imply 24 directions $\boldsymbol{d}_\alpha$ and 24 yield criteria f_α.
Since the dissipative stress ξ^{d} is defined by

$$\xi^{\mathrm{d}} = \xi - \xi^{\mathrm{e}} \overset{(4.17)}{=} \mathrm{div}\,(\boldsymbol{\xi}) - \partial_\zeta W, \tag{4.79}$$

an alternative representation of the yield criteria is given by

$$f_\alpha = \tau_\alpha + \operatorname{div}(\boldsymbol{\xi}) - (\tau_0^{\mathrm{C}} + \xi^{\mathrm{e}}). \tag{4.80}$$

Obviously, $\xi^{\mathrm{e}} = \partial_\zeta W_{\mathrm{h}}$ can be interpreted as classical isotropic hardening stress, and $\operatorname{div}(\boldsymbol{\xi}) = \operatorname{div}(\boldsymbol{\xi}^{\mathrm{e}})$ can be interpreted as an energetic gradient stress which reduces the inhomogeneity of the plastic deformation. The flow rule is assumed to be a superposition of the flow rules of the individual yield criteria (compare Eqns. (4.48) and (4.49))

$$\dot{\boldsymbol{\varepsilon}}^{\mathrm{p}} = \sum_\alpha \mathring{\lambda}_\alpha\, \partial_\sigma f_\alpha = \sum_\alpha \mathring{\lambda}_\alpha \boldsymbol{P}^{\mathrm{S}}_\alpha, \tag{4.81}$$

$$\dot{\zeta} = \sum_\alpha \mathring{\lambda}_\alpha\, \partial_{\xi^{\mathrm{d}}} f_\alpha = \sum_\alpha \mathring{\lambda}_\alpha. \tag{4.82}$$

This is in line with classical approaches involving several yield criteria (Koiter, 1953). The sums run over $\alpha = 1, ..., 2N$. The rate-dependency of the model is defined through the power law creep model

$$\mathring{\lambda}_\alpha = \dot{\gamma}_0 \left\langle \frac{f_\alpha}{\tau^{\mathrm{D}}} \right\rangle^p = \dot{\gamma}_0 \left\langle \frac{\tau_\alpha + \operatorname{div}(\boldsymbol{\xi}) - (\tau_0^{\mathrm{C}} + \xi^{\mathrm{e}})}{\tau^{\mathrm{D}}} \right\rangle^p, \tag{4.83}$$

which is in line with Eq. (4.52). Clearly, $\mathring{\lambda}_\alpha \geq 0$.
Let slip parameters be defined by

$$\lambda_\alpha = \int \mathring{\lambda}_\alpha \,\mathrm{d}t. \tag{4.84}$$

This definition implies $\dot{\lambda}_\alpha = \mathring{\lambda}_\alpha$.
The effective plastic shear strain of a given slip system is given by

$$\gamma_\alpha = \lambda_\alpha - \lambda'_\alpha. \tag{4.85}$$

Consequently, the plastic strain is given by

$$\boldsymbol{\varepsilon}^{\mathrm{p}} = \sum_\alpha \gamma_\alpha \boldsymbol{P}^{\mathrm{S}}_\alpha = \sum_\alpha \lambda_\alpha \boldsymbol{P}^{\mathrm{S}}_\alpha, \tag{4.86}$$

where the sums run over all slip systems and slip parameters, respectively. In many applications, an equivalent plastic shear strain is introduced by

$$\gamma_{\mathrm{eq}} = \sum_\alpha \int_0^t |\dot{\lambda}_\alpha| \,\mathrm{d}\tilde{t} = \sum_\alpha \lambda_\alpha. \tag{4.87}$$

Obviously, the plastic variable ζ coincides with the equivalent plastic strain (compare Eq. (4.82))

$$\gamma_{\mathrm{eq}} = \zeta. \tag{4.88}$$

Hence, the proposed gradient crystal plasticity theory coincides with the equivalent plastic strain gradient enhanced theory of Wulfinghoff and Böhlke (2012a), the micromorphic version of which is treated in Wulfinghoff and Böhlke (2013) and Wulfinghoff et al. (2013a) (see Chap. 5).

Chapter 5

Equivalent Plastic Strain Gradient Crystal Plasticity – Micromorphic Formulation and Implementation

In Chap. 4, a theoretical basis of phenomenological gradient crystal plasticity based on a scalar plastic field variable has been outlined. In the chapter at hand, this framework is discretized by the Finite Element Method and applied to micro-components. As a consequence, the presentation focuses on a formulation which is well-suited for numerical implementations. Therefore, a micromorphic setting is applied (Forest, 2009). In addition, the role of grain boundary models is emphasized. The content of the chapter has largely been taken from Wulfinghoff and Böhlke (2013) and Wulfinghoff et al. (2013a).

5.1 Theoretical Framework for Grain Boundary Resistance

5.1.1 Motivation

Many continuum mechanical single crystal models introduce an internal hardening variable which is often denoted as the equivalent (or accumulated) plastic strain $\gamma_{\rm eq}$. The equivalent plastic strain serves as a measure of the total plastic deformation at a material point. Specifically, it takes large values in plastically strongly deformed regions and small values in plastically weakly deformed regions.

The close formal relation of the gradient of the equivalent plastic strain $\nabla\gamma_{\rm eq}$ and Nye's dislocation density tensor (the curl of the plastic distorsion) serves as motivation for the interpretation of $\nabla\gamma_{\text{eq}}$ as approximate measure of geometrically necessary dislocations. Therefore, the gradient extension of the crystal plasticity theory at hand is established through $\nabla\gamma_{\text{eq}}$. The principal difference between both quantities is given by the fact that the gradient of the equivalent plastic strain $\nabla\gamma_{\text{eq}}$ is rather a measure of inhomogeneity than of incompatibility of the plastic deformation.

As has been outlined in Chap. 4, the main motivation for employing $\nabla\gamma_{\text{eq}}$ instead of the dislocation density tensor is given by its computational benefit, which involves the gradient of only one scalar field variable γ_{eq}. In contrast, the computation of the dislocation density tensor requires the evaluation of the spatial derivatives of all nine components of the plastic distorsion. As an additional motivation, the formulation of grain boundary models is expected to be conceptually more simple if only one scalar variable γ_{eq} is involved.

Of course, the theory cannot be expected to include the full richness of more complex strain gradient plasticity theories. For example, the shear bands observed in the work of Cordero et al. (2012) are not expected to be reproducible by the theory at hand since those are a result of the complex dislocation kinematics incorporated by the full dislocation density tensor. However, the model accounts for size effects concerning the material strength which are related to inhomogeneous plastic deformations of the model at the microscale.

As has been discussed by Wulfinghoff and Böhlke (2012a), in addition to γ_{eq}, a field variable ζ is needed for numerical reasons. It must be introduced in order to facilitate the aforementioned computational benefit. Without the introduction of ζ, the numerical implementation requires the evaluation of the gradients of all plastic slips, i.e., the computational merit of the approach would be lost (see Wulfinghoff and Böhlke, 2012a, for details). The equality of γ_{eq} and ζ has been established by the aforementioned authors in a weak sense by means of a Lagrange multiplier. In Chap. 4, the connection has been realized through the flow rule. For numerical reasons, this coupling is established in the chapter at hand by a penalty method which is in line with the micromorphic approach of Forest (2009). The gradient extension is then realized through ζ instead of γ_{eq}, which preserves its short-range character. As a result, the field $\zeta \approx \gamma_{\text{eq}}$ can be interpreted as the micromorphic counterpart of the equivalent plastic strain.

In this chapter, the equivalent plastic strain γ_{eq} and its micromorphic counterpart ζ remain close, since the micromorphic framework is applied like a penalty method. However, in general, the material parameters could also be chosen such that significant deviations between both quantities occur (Aslan et al., 2011; Poh et al., 2011). Therefore, the micromorphic variable ζ should be distinguished clearly from γ_{eq}.

5.1.2 Kinematical Assumptions

In the subsequent paragraphs a geometrically linear gradient plasticity framework is developed. The theory accounts for grain boundary resistance against plastic flow. A micromorphic theory is formulated, the kinematical framework of which differs from the one of Chap. 4. Therefore, the kinematics are shortly outlined. As before, material position vectors of the body $\mathcal{B}$ under consideration are referred to as $\boldsymbol{x}$, the displacements of the material points as $\boldsymbol{u}(\boldsymbol{x}, t)$ and the strain tensor is given by $\boldsymbol{\varepsilon} = \text{sym}(\nabla \boldsymbol{u})$, where $\nabla \boldsymbol{u} = \partial_j u_i \boldsymbol{e}_i \otimes \boldsymbol{e}_j$ in a Cartesian coordinate system with basis vectors $\{\boldsymbol{e}_1, \boldsymbol{e}_2, \boldsymbol{e}_3\}$. The plastic strain reads

$$\boldsymbol{\varepsilon}^{\text{p}} = \sum_{\alpha} \lambda_\alpha \boldsymbol{P}^{\text{S}}_\alpha, \tag{5.1}$$

where $\boldsymbol{P}^{\text{S}}_\alpha = \text{sym}(\boldsymbol{d}_\alpha \otimes \boldsymbol{n}_\alpha)$ and $\alpha = 1, \ldots, N$ are the slip indices. The directional dependent plastic slips are denoted by λ_α, the slip directions by $\boldsymbol{d}_\alpha$, and the slip plane normals by $\boldsymbol{n}_\alpha$. Each slip system is defined by its slip plane and opposite slip direction pairs. Furthermore, it is required that the directional dependent plastic slips increase monotonously, i.e. $\dot{\lambda}_\alpha \geq 0$. Consequently, the total plastic slip of each slip system is given by the difference between its two directional dependent plastic slips. For a face centered cubic (FCC) crystal this implies $N = 24$.

The elastic part of the strain tensor ε is defined by $\varepsilon^{\mathrm{e}} = \varepsilon - \varepsilon^{\mathrm{p}}$. In addition, the equivalent plastic strain is introduced. Here, it takes the form

$$\gamma_{\mathrm{eq}}(\hat{\lambda}) = \sum_{\alpha} \int_0^t |\dot{\lambda}_\alpha| \, \mathrm{d}\tilde{t} = \sum_{\alpha} \lambda_\alpha. \tag{5.2}$$

5.1.3 Principle of Virtual Power and Field Equations

The principle of virtual power follows the same lines as in Sect. 4.3. For completeness, it is shortly reviewed. Neglecting body forces, the virtual power of the external forces is assumed to be given by

$$\delta \mathcal{P}_{\mathrm{ext}} = \int_{\partial \mathcal{B}_t} \bar{\boldsymbol{t}} \cdot \delta \dot{\boldsymbol{u}} \, \mathrm{d}a + \int_{\partial \mathcal{B}_\Xi} \bar{\Xi} \, \delta \dot{\zeta} \, \mathrm{d}a. \tag{5.3}$$

The virtual power of the internal forces is assumed to be given by

$$\delta \mathcal{P}_{\mathrm{int}} = \int_{\mathcal{B}} \delta p_{\mathrm{int}} \, \mathrm{d}v + \int_{\Gamma} \Xi_\Gamma \, \delta \dot{\zeta} \, \mathrm{d}a, \tag{5.4}$$

where Γ denotes the union of all grain boundaries. The internal virtual power density is assumed to be given by

$$\delta p_{\mathrm{int}} = \boldsymbol{\sigma} \cdot \delta \dot{\boldsymbol{\varepsilon}} + \xi \, \delta \dot{\zeta} + \boldsymbol{\xi} \cdot \nabla \delta \dot{\zeta}. \tag{5.5}$$

The stresses $\boldsymbol{\sigma}$, ξ and $\boldsymbol{\xi}$ are work conjugate to the associated kinematical quantities, i.e., the associated power density of the internal forces,

$$p_{\mathrm{int}} = \boldsymbol{\sigma} \cdot \dot{\boldsymbol{\varepsilon}} + \xi \, \dot{\zeta} + \boldsymbol{\xi} \cdot \nabla \dot{\zeta}, \tag{5.6}$$

may be stored energetically or dissipated at the material point under consideration. In an analogous manner, the grain boundary microtraction Ξ_Γ is conjugate to $\dot{\zeta}$ on Γ. Possible jumps $[\![\zeta]\!]$ at the grain boundary are neglected in order to keep the number of necessary constitutive equations and new material parameters low.

The field equations as well as the Neumann and the grain boundary conditions can be obtained from the principle of virtual power by a standard procedure (see Sect. 4.3). They are summarized in Box 1.

Linear momentum	$\mathbf{0}$	$= \mathrm{div}\,(\boldsymbol{\sigma})$	$\forall \boldsymbol{x} \in \mathcal{B}$
Microforce balance	ξ	$= \mathrm{div}\,(\boldsymbol{\xi})$	$\forall \boldsymbol{x} \in \mathcal{B} \setminus \Gamma$
GB microtraction	Ξ_Γ	$= [\![\boldsymbol{\xi}]\!] \cdot \boldsymbol{n}$	$\forall \boldsymbol{x} \in \Gamma$
Neumann BCs	$\boldsymbol{\sigma n}$	$= \bar{\boldsymbol{t}}\ \forall \boldsymbol{x} \in \partial\mathcal{B}_t,$	$\boldsymbol{\xi} \cdot \boldsymbol{n} = \bar{\Xi}\ \forall \boldsymbol{x} \in \partial\mathcal{B}_\Xi$

Box 1: Field equations, Neumann and grain boundary conditions. The grain boundary jump is denoted by $[\![\boldsymbol{\xi}]\!] = \boldsymbol{\xi}^+ - \boldsymbol{\xi}^-$ with the normal $\boldsymbol{n}$ pointing from '-' to '+'.

The following relation is obtained by the choice $\delta\dot{\boldsymbol{u}} = \dot{\boldsymbol{u}}$ and $\delta\dot{\zeta} = \dot{\zeta}$

$$\mathcal{P}_{\text{int}} = \mathcal{P}_{\text{ext}}, \tag{5.7}$$

i.e., the power of the internal forces equals the power of the external forces. However, note that the integral represented by $\mathcal{P}_{\text{ext}}$ takes into account the whole boundary $\partial\mathcal{B}$, contrary to $\delta\mathcal{P}_{\text{ext}}$ (cf. Eq. (5.3)).

5.1.4 Constitutive Assumptions

Stored Energy Density

The stored energy density is assumed to have the following contributions

$$W(\boldsymbol{\varepsilon}, \hat{\lambda}, \zeta, \nabla\zeta) = W_{\text{e}}(\boldsymbol{\varepsilon}, \boldsymbol{\varepsilon}^{\text{p}}(\hat{\lambda})) + W_{\text{h}}(\zeta) + W_{\text{g}}(\nabla\zeta) + W_{\chi}(\zeta - \gamma_{\text{eq}}(\hat{\lambda})), \tag{5.8}$$

with $W_{\text{e}}(\boldsymbol{\varepsilon}, \boldsymbol{\varepsilon}^{\text{p}}) = 1/2(\boldsymbol{\varepsilon} - \boldsymbol{\varepsilon}^{\text{p}}) \cdot \mathbb{C}[\boldsymbol{\varepsilon} - \boldsymbol{\varepsilon}^{\text{p}}]$, where $\mathbb{C}$ denotes the elastic stiffness tensor. Deviations of ζ from γ_{eq} are penalized by the energy $W_\chi = H_\chi(\zeta - \gamma_{\text{eq}})^2/2$ by use of a large penalty factor H_χ. Hardening mechanisms due to dislocation multiplication and trapping are modeled through the phenomenological isotropic hardening energy $W_{\text{h}}(\zeta)$. The gradient hardening energy $W_{\text{g}}(\nabla\zeta)$ introduces an internal length scale into the theory which plays a significant role in the context of size effects (see Chap. 4). Additionally, an interface energy per unit surface $W_\Gamma(\zeta)$ is introduced on the grain boundaries Γ in order to account for energetical grain boundary resistance against plastic flow. Note that the penalty energy W_χ couples the two field variables $\gamma_{\rm eq}$ and ζ. In this sense, the model contrasts with the theory outlined in Sect. 4.6. As a consequence, the dissipation inequality changes, as is shown subsequently.

Dissipation Inequality

In case of a purely mechanical theory thermal effects are neglected and the total dissipation $D_{\rm tot}$ is given by

$$D_{\rm tot} = \mathcal{P}_{\rm ext} - \int_{\mathcal{B}} \dot{W} \, {\rm d}v - \int_{\Gamma} \dot{W}_\Gamma \, {\rm d}a \geq 0. \tag{5.9}$$

From $\mathcal{P}_{\rm int} = \mathcal{P}_{\rm ext}$ (Eq. (5.7)) it follows that

$$D_{\rm tot} = \int_{\mathcal{B}} \mathcal{D} \, {\rm d}v + \int_{\Gamma} \mathcal{D}_\Gamma \, {\rm d}a \geq 0, \tag{5.10}$$

where the bulk dissipation is given by (cf. Eqns. (5.6) and (5.8))

$$\begin{aligned} \mathcal{D} &= (\boldsymbol{\sigma} - \partial_{\boldsymbol{\varepsilon}} W_{\rm e}) \cdot \dot{\boldsymbol{\varepsilon}} - \partial_{\boldsymbol{\varepsilon}^{\rm p}} W_{\rm e} \cdot \dot{\boldsymbol{\varepsilon}}^{\rm p} - \partial_{\gamma_{\rm eq}} W_\chi \dot{\gamma}_{\rm eq} & (5.11) \\ &+ (\xi - \partial_\zeta W_{\rm h} - \partial_\zeta W_\chi) \, \dot{\zeta} + (\boldsymbol{\xi} - \partial_{\nabla\zeta} W_{\rm g}) \cdot \nabla \dot{\zeta} \geq 0. & (5.12) \end{aligned}$$

At the grain boundary, the dissipation (per unit area) reads

$$\mathcal{D}_\Gamma = (\Xi_\Gamma - \partial_\zeta W_\Gamma) \, \dot{\zeta} = (\Xi_\Gamma - \Xi_\Gamma^{\rm e}) \, \dot{\zeta} = \Xi_\Gamma^{\rm d} \, \dot{\zeta} \geq 0. \tag{5.13}$$

Here, the energetic and dissipative grain boundary microtractions $\Xi_\Gamma^{\rm e}$ and $\Xi_\Gamma^{\rm d}$ have been introduced

$$\Xi_\Gamma^{\rm e} = \partial_\zeta W_\Gamma, \quad \Xi_\Gamma^{\rm d} = \Xi_\Gamma - \Xi_\Gamma^{\rm e}. \tag{5.14}$$

For better readability, the following notation is introduced

$$\beta = \partial_\zeta W_{\rm h}, \quad \check{p} = -\partial_\zeta W_\chi = \partial_{\gamma_{\rm eq}} W_\chi = H_\chi (\gamma_{\rm eq} - \zeta). \tag{5.15}$$

It is assumed, that the stresses $\boldsymbol{\sigma}$, ξ and $\boldsymbol{\xi}$ are purely energetic, i.e.

$$\boldsymbol{\sigma} = \partial_{\boldsymbol{\varepsilon}} W_{\rm e}, \quad \xi = \partial_\zeta W = \beta - \check{p}, \quad \boldsymbol{\xi} = \partial_{\nabla\zeta} W_{\rm g}, \tag{5.16}$$

i.e., possible dissipative stresses are neglected.

These assumptions lead to the reduced dissipation inequality

$$\mathcal{D} = -\partial_{\varepsilon^{\mathrm{p}}} W_{\mathrm{e}} \cdot \dot{\varepsilon}^{\mathrm{p}} - \partial_{\gamma_{\mathrm{eq}}} W_{\chi} \dot{\gamma}_{\mathrm{eq}} = \sum_{\alpha} (\tau_\alpha - \check{p})\, \dot{\lambda}_\alpha \geq 0. \qquad (5.17)$$

Here, the relations $\boldsymbol{\sigma} = -\partial_{\varepsilon^{\mathrm{p}}} W_{\mathrm{e}}$ and Eqns. (5.1), (5.2) and (5.15) have been exploited. The resolved shear stresses are defined by $\tau_\alpha = \boldsymbol{\sigma} \cdot \boldsymbol{P}^{\mathrm{S}}_\alpha$.

Flow Rules

In order to be thermodynamically consistent, the bulk flow rule must satisfy Eq. (5.17). A possible choice is given by the power law type creep law

$$\dot{\lambda}_\alpha = \dot{\gamma}_0 \left\langle \frac{\tau_\alpha - \check{p} - \tau_0^{\mathrm{C}}}{\tau^{\mathrm{D}}} \right\rangle^p = \dot{\gamma}_0 \left\langle \frac{\tau_\alpha + \operatorname{div}(\boldsymbol{\xi}) - (\tau_0^{\mathrm{C}} + \beta)}{\tau^{\mathrm{D}}} \right\rangle^p. \qquad (5.18)$$

Here, the microforce balance (Box 1) and Eq. (5.16) have been exploited. The reference shear rate $\dot{\gamma}_0$, the drag stress τ^{D}, the initial yield stress τ_0^{C} and the strain rate sensitivity p are material parameters. Isotropic hardening is accounted for through the stress $\beta = \partial_\zeta W_{\mathrm{h}}$. Formally, the term $\operatorname{div}(\boldsymbol{\xi})$ is related to more complex gradient theories (e.g. Gurtin et al., 2007), where it has the interpretation of a backstress. However, the kinematic hardening effects observed in those theories are in general not expected here. In fact, the term $\operatorname{div}(\boldsymbol{\xi})$ rather leads to a reduction of inhomogeneity of the plastic deformation, since it is a direct consequence of the introduction of $W_{\mathrm{g}}(\nabla\zeta)$. This implies, in general, the effect that the smaller the system the stronger is the mechanical response to inhomogeneous deformations.

The grain boundary dissipation inequality (5.13) motivates the following grain boundary yield condition (cf. Box 1 and Eq. (5.14))

$$f_\Gamma = \Xi^{\mathrm{d}}_\Gamma - \Xi^{\mathrm{C}}_0 = [\![\boldsymbol{\xi}]\!] \cdot \boldsymbol{n} - (\Xi^{\mathrm{C}}_0 + \Xi^{\mathrm{e}}_\Gamma), \qquad (5.19)$$

where Ξ^{C}_0 is the dissipative contribution to the yield strength, which is assumed to be constant. In the case $f_\Gamma < 0 \rightarrow \dot{\zeta} = 0$ holds. Assuming a rate-independent grain boundary model, grain boundary yielding occurs if the yield criterion $f_\Gamma = 0$ and the loading condition $\dot{f}_\Gamma\big|_\zeta > 0 \to \dot{\zeta} > 0$ is fulfilled. Thus, the Kuhn-Tucker conditions for the grain boundaries can be expressed by

$$f_\Gamma \leq 0, \quad \dot{\zeta} \geq 0, \quad \dot{\zeta} f_\Gamma = 0. \tag{5.20}$$

Microfree and microhard grain boundaries are represented by the two limit cases $\Xi_0^C = 0,\ W_\Gamma = 0$ and, e.g., $\Xi_0^C \to \infty$, respectively. The grain boundary dissipation (per unit area) reads $\mathcal{D}_\Gamma = \Xi_0^C \dot{\zeta}$ (cf. Eqns. (5.13) and (5.20)). It should be noted that a rate-dependent grain boundary flow rule is also possible.

5.2 Finite Element Implementation

5.2.1 Linearization of the Variational Form

The finite element implementation of the model involves the computation of the field variables $\boldsymbol{u}$ and ζ, which are solved for on a global level by a classical Newton algorithm. This means that each node has 4 degrees of freedom $\{\boldsymbol{u}, \zeta\}$. The stresses and internal variables are computed in each global Newton step by a separate integration point algorithm. The consistent linearization of this local algorithm is required in order to compute the global stiffness matrix for the aforementioned Newton scheme. The basis for the global Newton algorithm is the linearization of the principle of virtual power (Eqns. (5.3) and (5.4)). Time steps are addressed by subscripts, e.g. "n", in the following. For simplicity the subscript "$\mathrm{n}+1$" is dropped for quantities of a subsequent time step, i.e. $\tau_\alpha = \tau_{\alpha,\mathrm{n}+1}$, for example.
The linearization of the terms in $\delta\dot{\boldsymbol{u}}$ yields

$$\int_{\mathcal{B}} (\partial_\varepsilon \boldsymbol{\sigma}[\Delta\boldsymbol{\varepsilon}] + \partial_\zeta \boldsymbol{\sigma} \Delta\zeta) \cdot \delta\dot{\boldsymbol{\varepsilon}} \,\mathrm{d}v = -\int_{\mathcal{B}} \boldsymbol{\sigma} \cdot \delta\dot{\boldsymbol{\varepsilon}} \,\mathrm{d}v + \int_{\partial\mathcal{B}_t} \bar{\boldsymbol{t}} \cdot \delta\dot{\boldsymbol{u}} \,\mathrm{d}a \quad \forall \delta\dot{\boldsymbol{u}}. \tag{5.21}$$

The partial derivatives denote the algorithmic tangents. Furthermore, the linearization of the terms in $\delta\dot{\zeta}$ reads

$$\int_{\mathcal{B}} \Big(\delta\dot{\zeta}\big(\partial_\zeta \beta - \partial_\zeta \check{p}\big)\Delta\zeta - \delta\dot{\zeta}\,\partial_{\varepsilon}\check{p}\cdot\Delta\varepsilon + \big(\partial_{\nabla\zeta}\boldsymbol{\xi}\,\nabla(\Delta\zeta)\big)\cdot\nabla(\delta\dot{\zeta})\Big)\,\mathrm{d}v$$

$$+\int_{\Gamma_{\mathrm{act}}} \delta\dot{\zeta}\,\partial_\zeta\,\Xi^{\mathrm{e}}_{\Gamma}\,\Delta\zeta\,\mathrm{d}a = -\int_{\mathcal{B}}\Big(\big(\beta-\check{p}\big)\delta\dot{\zeta} + \boldsymbol{\xi}\cdot\nabla(\delta\dot{\zeta})\Big)\,\mathrm{d}v + \int_{\partial\mathcal{B}_{\Xi}} \bar{\Xi}\,\delta\dot{\zeta}\,\mathrm{d}a$$

$$-\underbrace{\int_{\Gamma\backslash\Gamma_{\mathrm{act}}} (\Xi^{\mathrm{d}}_{\Gamma}+\Xi^{\mathrm{e}}_{\Gamma})\,\delta\dot{\zeta}\,\mathrm{d}a}_{I} - \int_{\Gamma_{\mathrm{act}}} (\Xi^{\mathrm{C}}_{0}+\Xi^{\mathrm{e}}_{\Gamma})\,\delta\dot{\zeta}\,\mathrm{d}a, \tag{5.22}$$

where it is exploited that the yield condition $f_\Gamma = \Xi^{\mathrm{d}}_{\Gamma} - \Xi^{\mathrm{C}}_{0} = 0$ is fulfilled on $\Gamma_{\mathrm{act}} = \{\boldsymbol{x}\in\Gamma : \dot{\zeta} > 0\}$. It should also be noted that the integral I is irrelevant for the computation if $\delta\dot{\zeta}$ is chosen to vanish on $\Gamma\backslash\Gamma_{\mathrm{act}}$ (for the same reason the linearization of I is neglected).
Equations (5.21) and (5.22) are discretized in space with the Finite Element Method. The numerical integration of Eqns. (5.21) and (5.22) requires the computation of the different stresses and tangents by the integration point routine.

5.2.2 Numerical Time Integration

In the following, the computation of the stresses in Eqns. (5.21) and (5.22) is shortly summarized. This computation is effectuated by an integration point subroutine during each Newton iteration. The implicit Euler scheme is applied to Eq. (5.18)

$$\lambda_\alpha = \lambda_{\alpha,\mathrm{n}} + \dot{\gamma}_0\Delta t\left\langle\frac{\boldsymbol{\sigma}\cdot\boldsymbol{P}^{\mathrm{S}}_{\alpha} - \check{p} - \tau^{\mathrm{C}}_{0}}{\tau^{\mathrm{D}}}\right\rangle^{p}. \tag{5.23}$$

The first equation of the local Newton scheme reads

$$\boldsymbol{r}^{\sigma} = -\mathbb{S}[\boldsymbol{\sigma}] + \boldsymbol{\varepsilon} - \boldsymbol{\varepsilon}^{\mathrm{p}}_{\mathrm{n}} - \sum_{\alpha}\Delta t\dot{\gamma}_0\left\langle\frac{\boldsymbol{\sigma}\cdot\boldsymbol{P}^{\mathrm{S}}_{\alpha} - \check{p} - \tau^{\mathrm{C}}_{0}}{\tau^{\mathrm{D}}}\right\rangle^{p}\boldsymbol{P}^{\mathrm{S}}_{\alpha} = \mathbf{0}, \tag{5.24}$$

with $\mathbb{S} = \mathbb{C}^{-1}$. This equation states $\boldsymbol{\varepsilon} = \boldsymbol{\varepsilon}^{\mathrm{e}} + \boldsymbol{\varepsilon}^{\mathrm{p}}$, since $\mathbb{S}[\boldsymbol{\sigma}] = \boldsymbol{\varepsilon}^{\mathrm{e}}$ and $\sum_\alpha \Delta t\dot{\gamma}_0\left\langle(\tau_\alpha - \check{p} - \tau^{\mathrm{C}}_{0})/\tau^{\mathrm{D}}\right\rangle^{p}\boldsymbol{P}^{\mathrm{S}}_{\alpha} = \Delta\boldsymbol{\varepsilon}^{\mathrm{p}}$.

The second part of the residual can be deduced from Eq. (5.15)

$$r^{\mathrm{p}} = \underbrace{\gamma_{\mathrm{eq,n}} + \sum_{\alpha} \Delta t \dot{\gamma}_0 \left\langle \frac{\boldsymbol{\sigma} \cdot \boldsymbol{P}^{\mathrm{S}}_{\alpha} - \check{p} - \tau^{\mathrm{C}}_0}{\tau^{\mathrm{D}}} \right\rangle^{p}}_{\Delta\gamma_{\mathrm{eq}}} - \zeta - \frac{\check{p}}{H_{\chi}} = 0. \tag{5.25}$$

Equations (5.24) and (5.25) are solved for $\boldsymbol{\sigma}$ and $\check{p}$ by an enhanced Newton algorithm (see Sect. 5.4).

5.2.3 Algorithmic Tangent

The application of Newton's scheme to the global problem requires the computation of the algorithmic tangent operators which appear in Eqns. (5.21) and (5.22). This must be effectuated at each integration point of the mesh in each global Newton iteration. In the following $\mathrm{d}\boldsymbol{r}^{\sigma} = \boldsymbol{0}$, and $\mathrm{d}r^{\mathrm{p}} = 0$ as a consequence of any perturbation in the global fields $\boldsymbol{u}$ and ζ in order to obtain the algorithmic tangent. The total differentials of the residuals are then given by

$$\mathrm{d}\boldsymbol{r}^{\sigma} = \partial_{\sigma}\boldsymbol{r}^{\sigma}[\mathrm{d}\boldsymbol{\sigma}] + \mathbb{I}^{\mathrm{s}}[\mathrm{d}\boldsymbol{\varepsilon}] + \partial_{\check{p}}\boldsymbol{r}^{\sigma}\,\mathrm{d}\check{p} = \boldsymbol{0}, \tag{5.26}$$

and

$$\mathrm{d}r^{\mathrm{p}} = \partial_{\sigma} r^{\mathrm{p}} \cdot \mathrm{d}\boldsymbol{\sigma} + \partial_{\check{p}}\, r^{\mathrm{p}}\,\mathrm{d}\check{p} + \partial_{\zeta}\, r^{\mathrm{p}}\,\mathrm{d}\zeta = 0. \tag{5.27}$$

This can be rewritten in matrix-vector notation by

$$\underbrace{\begin{pmatrix} -\partial_{\hat{\sigma}}\,\hat{r}^{\sigma} & -\partial_{\check{p}}\,\hat{r}^{\sigma} \\ -\partial_{\hat{\sigma}}\,r^{\mathrm{p}} & -\partial_{\check{p}}\,r^{\mathrm{p}} \end{pmatrix}}_{\hat{A}} \begin{pmatrix} \mathrm{d}\hat{\sigma} \\ \mathrm{d}\check{p} \end{pmatrix} = \underbrace{\begin{pmatrix} \hat{I} & 0 \\ 0 & -1 \end{pmatrix}}_{\hat{B}} \begin{pmatrix} \mathrm{d}\hat{\varepsilon} \\ \mathrm{d}\zeta \end{pmatrix}, \tag{5.28}$$

and thus

$$\begin{pmatrix} \mathrm{d}\hat{\sigma} \\ \mathrm{d}\check{p} \end{pmatrix} = \underbrace{\hat{A}^{-1}\hat{B}}_{\hat{D}} \begin{pmatrix} \mathrm{d}\hat{\varepsilon} \\ \mathrm{d}\zeta \end{pmatrix}, \tag{5.29}$$

where

$$\hat{D} = \begin{pmatrix} \partial_{\hat{\varepsilon}}\,\hat{\sigma} & \partial_{\zeta}\,\hat{\sigma} \\ \partial_{\hat{\varepsilon}}\,\check{p} & \partial_{\zeta}\,\check{p} \end{pmatrix}, \tag{5.30}$$

with $\hat{D} \in Sym$ due to $\partial_{\sigma}\boldsymbol{r}^{\sigma} \in Sym$ and $\partial_{\check{p}}\boldsymbol{r}^{\sigma} = -\partial_{\sigma} r^{\mathrm{p}}$.

5.2.4 Discretization of the Grain Boundaries

Equations (5.21) and (5.22) are discretized by the standard Finite Element Method except for the integral over Γ_{act} in Eq. (5.22) which is approximated by

$$-\int_{\Gamma_{\text{act}}} (\Xi_0^{\text{C}} + \Xi_\Gamma^{\text{e}}) \delta\dot{\zeta} \, \text{d}a \cong - \sum_{i \in \Gamma_{\text{act}}} (\Xi_0^{\text{C}} + \Xi_\Gamma^{\text{e}})_i w_i \delta\dot{\zeta}_i, \tag{5.31}$$

with the integration points i and the quadrature weights w_i. The integration points are chosen to be coinciding with the nodes of the Finite Element mesh. This strategy, which is inspired by the work of Liebe and Steinmann (2001), eases the determination of the plastically active part Γ_{act} of the grain boundaries and is well-established for linear elements. The discretized counterpart of Γ_{act} is given by the active set $\mathcal{A}$ of grain boundary nodes where $f_\Gamma = 0$ and $\Delta\zeta \geq 0$ (see Eq. (5.20)). Inactive grain boundary nodes are treated as Dirichlet nodes with $\Delta\zeta = 0$ and $\delta\dot{\zeta} = 0$. Hence, the explicit computation of the integral

$$-\int_{\Gamma \backslash \Gamma_{\text{act}}} (\Xi_\Gamma^{\text{d}} + \Xi_\Gamma^{\text{e}}) \, \delta\dot{\zeta} \, \text{d}a \cong - \sum_{i \in \Gamma \backslash \Gamma_{\text{act}}} (\Xi_\Gamma^{\text{d}} + \Xi_\Gamma^{\text{e}})_i w_i \delta\dot{\zeta}_i, \tag{5.32}$$

in Eq. (5.22) is not necessary (and not possible), since the associated lines and columns are eliminated from the global stiffness matrix and the residual, respectively. Due to the neglect of Eq. (5.32) during the assembly, the corresponding components of the force residual do not become zero when convergence is reached but take the values $(\Xi_\Gamma^{\text{d}} + \Xi_\Gamma^{\text{e}})_i w_i$. This result is exploited to evaluate the yield criterion $f_\Gamma = \Xi_\Gamma^{\text{d}} - \Xi_0^{\text{C}}$ at the inactive nodes and eventually update the active set $\mathcal{A}$.

The full algorithmic determination of $\mathcal{A}$ as well as the contribution of the grain boundary term in Eq. (5.31) to the global stiffness matrix and force residual are presented in Box 2.

1. Grain boundary contribution to the residual and the stiffness matrix:
   For all grain boundary nodes in active set $\mathcal{A}$:
   a) Add $-(\Xi_\Gamma^e + \Xi_0^C)_i w_i$ to the right-hand side of the global force residual (cf. Eqns. (5.22) and (5.31))
   b) Add $(\partial_{\zeta\zeta} W_\Gamma)_i w_i$ to the diagonal of the global stiffness matrix (If W_Γ is assumed to be linear in ζ the stiffness matrix consequently does not need to be updated)
2. Update of active set $\mathcal{A}$:
   If res* $<$ tol$_\mathcal{A}$ (i.e. update $\mathcal{A}$ only if close to convergence)
   a) For all grain boundary nodes i:
      If $i \in \mathcal{A}$

      If $\Delta\zeta_i = \zeta_i - \zeta_{i,\mathrm{n}} < 0$
      - $\mathcal{A} \leftarrow \mathcal{A} \backslash \{i\}$
      - $\zeta_i = \zeta_{i,\mathrm{n}}$

      Else

      If $f_\Gamma = \Xi_\Gamma^d - \Xi_0^C > 0$
      - $\mathcal{A} \leftarrow \mathcal{A} \cup \{i\}$
   b) If active set $\mathcal{A}$ has changed:
      Recompute the stiffness matrix and the residuals (including step 1)
3. If res $<$ tol$_\mathrm{force}$ and $\mathcal{A}$ has not changed the time step is considered converged

* res is the maximum norm of the force residual

Box 2: Algorithm for determination of plastic activity of grain boundary nodes.

5.3 Numerical Results

5.3.1 Simulation Setup and Boundary Conditions

In this section at first some general details of the simulations are presented. Subsequently, the discretization and the model parameters used are discussed, with emphasis on the influence of the introduced grain boundary yield criterion. Finally, the numerical results are compared to experimental results from the literature.
Throughout the simulations, the considered cubic oligocrystals consist of 2, 8 and 64 grains, respectively, with simplified geometries and randomly assigned crystal orientations. The length of the cubic structures is 25 µm each, i.e., for an 8 grain aggregate, e.g., each grain is represented by a cubic 12.5 x 12.5 x 12.5 μm^3 volume. The oligocrystals are loaded in tension. It should be noted that free lateral contraction is permitted by the chosen boundary conditions throughout the simulations in this work. In detail, the boundary conditions for the displacements in x, y, z-direction, respectively, are depicted in Fig. 5.1.
All initial values are set to zero. This means that the body is completely undeformed at the beginning of the first time increment. The whole boundary $\partial\mathcal{B}$ is treated as Neumann boundary with $\bar{\Xi} = 0$. All simulations are performed using the in-house Finite Element code of the Institute of Engineering Mechanics (Chair for Continuum Mechanics).

5.3.2 Discretization Details

Standard linear hexahedrons are used for all discretizations. The number of elements is 64000 in all presented simulation results (element length $h = 25\mu m/40$), cf. Fig. 5.2, if not indicated contrarily. At each node 4 degrees of freedom (DOFs), i.e., three for the displacements $\boldsymbol{u}$ and one for the field ζ, exist. An adaptive time stepping algorithm is

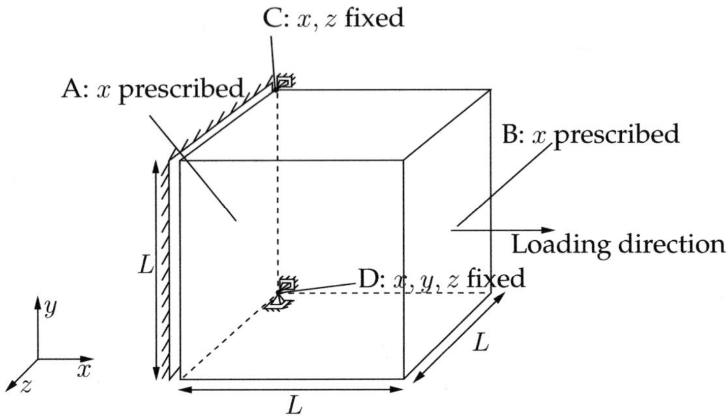

Figure 5.1: Tensile test FEM displacement boundary conditions. On plane A the displacement in x-direction is prescribed to remain zero, and on plane B the loading is prescribed via the displacement in x-direction. Furthermore, at node C the displacement is fixed for x- and z-direction as well as for x-, y- and z-direction at node D, respectively. The length is $L = 25$ μm.

used during the total simulation time of 1 s. The global Newton scheme is considered to be converged, if the initial residual is reduced by a factor of 10^{-8} (i.e. $\mathrm{tol}_{\mathrm{force}} = 10^{-8}\,\mathrm{res}_0$, see Box 2) and if the set of active grain boundary nodes $\mathcal{A}$ is not changed during an iteration. The tolerance for the active set search is taken as $\mathrm{tol}_{\mathcal{A}} = 10\,\mathrm{tol}_{\mathrm{force}}$ (see Box 2). However, an absolute tolerance of 10^{-8} (associated to the maximum norm of the local residual, Eqns. (5.24) and (5.25)) is applied for the integration point subroutine.

Residuals for two characteristic timesteps are listed in Tab. 5.1. The need for the active set search on the grain boundaries leads to additional iterations, mainly in the elastic-plastic transition regime. However, once all grain boundary nodes are plastically active, these additional iterations do not further occur.

The convergence of the results is visualized in Fig. 5.3. As a fair compromise of computational time and accuracy, the mesh with 275684 degrees of freedom is used for all presented simulations.

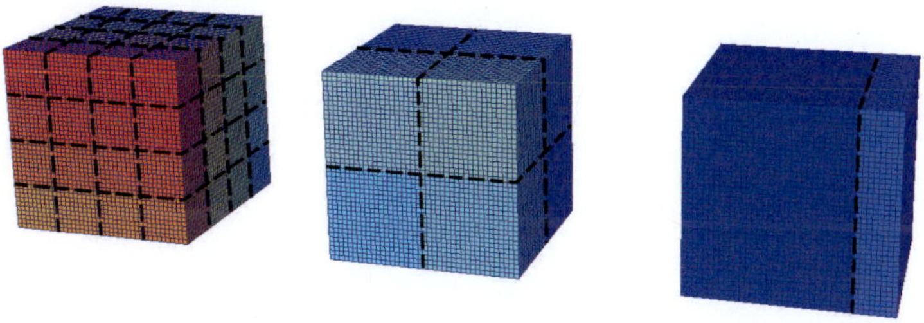

Figure 5.2: Discretizations of simplified grain aggregates used for simulating the mechanical response of annealed copper microwire tensile tests from Yang et al. (2012). From left to right: Simplified grain aggregate used for microwires annealed at 400°C, 600°C, and 900°C. Grain boundaries are highlighted by dashed lines.

Table 5.1: Residuals for two typical timesteps, one with active set search (in the elastic-plastic transition regime) and one without active set search (in the plastic regime). The highlighted residual values indicate iterations after which an update of the active set $\mathcal{A}$ occurs. The data is taken from a simulation of an aggregate of 64 crystals.

Δt res_0	Maximum norm of residual								
$0.01s$	Iteration 1	2	3	4	5	6	7	8	9
1.84e+03	5.67e+00	6.70e+00	7.96e-01	9.90e-02	1.75e-02	4.62e-04	**5.93e-07**	6.69e+00	1.15e+00
	10	11	**12**	13	14	**15**	16	17	
	1.20e-01	3.09e-03	**5.11e-06**	1.34e+00	1.25e-02	**2.99e-06**	1.14e-03	2.24e-09	
$0.32s$	Iteration 1	2	3	4	5	6	7		
5.88e+04	1.52e+01	9.92e+00	2.83e+00	4.92e-01	3.60e-02	1.21e-03	5.29e-06		

5.3.3 Model Parameters

For the contributions to the stored energy, cf. Eq. (5.8), a quadratic defect energy approach

$$W_{\text{g}}(\nabla\zeta) = \frac{1}{2} K_{\text{G}} \nabla\zeta \cdot \nabla\zeta \tag{5.33}$$

and a Voce hardening relation

$$W_{\text{h}}(\zeta) = (\tau_\infty^{\text{C}} - \tau_0^{\text{C}})\zeta + \frac{1}{\Theta_0}(\tau_\infty^{\text{C}} - \tau_0^{\text{C}})^2 \exp\left(-\frac{\Theta_0 \zeta}{\tau_\infty^{\text{C}} - \tau_0^{\text{C}}}\right), \tag{5.34}$$

are considered. A material length scale parameter l is implicitly introduced through $K_{\text{G}} = l^2 E$ (where E denotes Young's modulus of the

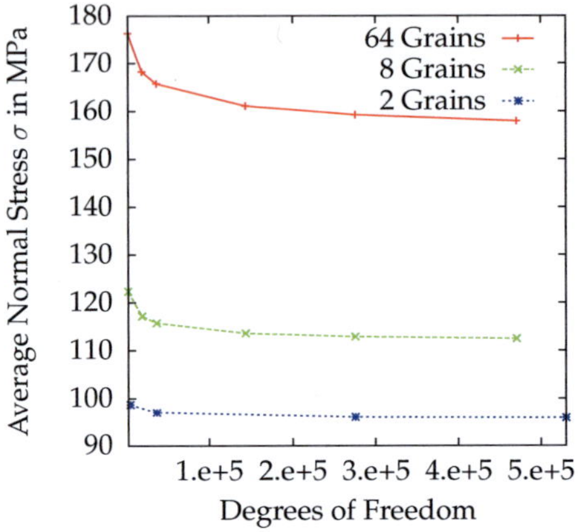

Figure 5.3: Final average normal stress at the end of the last time increment versus degrees of freedom for oligocrystals under tensile load with 64, 8 and 2 grains, respectively, using different discretizations. The diminution of DOFs due to the Dirichlet boundary conditions is neglected.

macroscopic material). The grain boundary energy W_Γ is set to zero. For simplicity, the grain boundary yield strength Ξ_0^C is assumed to be uniform and constant on the grain boundaries. Since the grain boundary yield strength does not depend on the crystal orientation in the model at hand, effects regarding the grain boundary resistance resulting from the orientations are not considered in the simulations. Consequently, the grain boundary yield criterion is given by $f_\Gamma = [\![\boldsymbol{\xi}]\!] \cdot \boldsymbol{n} - \Xi_0^C$. It should be noted, however, that instead of the dissipative treatment, the same mechanical model behavior can be achieved with the introduction of a linear grain boundary energy $W_\Gamma = \Xi_0^C \zeta$.

Exemplarily, the influence of Ξ_0^C and K_G (while keeping all other parameters constant) on the mechanical response of an oligocrystal with 8 grains under tensile load is illustrated in Fig. 5.4.

Upon an increase of Ξ_0^C the material responds with an increased yield strength in the region of well established plastic flow, leading to a stiffer behavior, cf. Fig. 5.4. As depicted, K_G mainly influences the elastic-plastic transition regime and the overall hardening. Furthermore, the used model parameters for the simulations are listed in Tab. 5.2, except for Ξ_0^{C} and K_{G}.

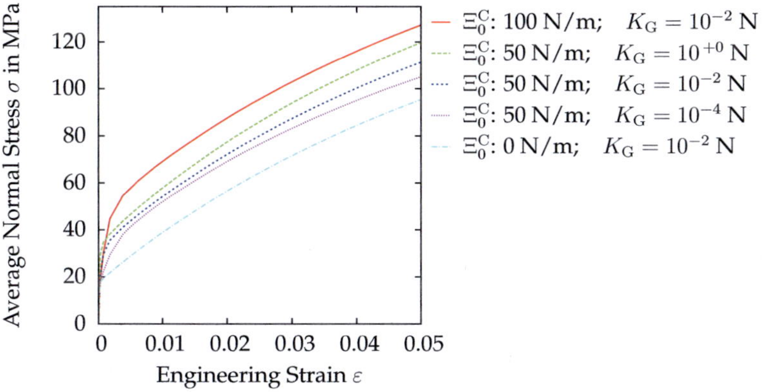

Figure 5.4: Influence of Ξ_0^{C} and K_{G}, respectively, on the mechanical response of a 25 µm oligocrystal with 8 grains under tensile load.

Table 5.2: Model parameters for simulations of copper oligocrystals, fitted to experimental data of Yang et al. (2012).

C_{1111}	C_{1122}	C_{1212}	H_χ	K_{G}	p
168 GPa	121 GPa	75 GPa	10^6 MPa	10^{-2} N	20
$\dot{\gamma}_0$	Θ_0	Ξ_0^{C}	τ_0^{C}	τ_∞^{C}	τ^{D}
10^{-3} 1/s	330 MPa	55 N/m	6 MPa	55 MPa	1 MPa

It should be noted that the depicted simulation results for $K_{\mathrm{G}} = 10^{-4}$ N can only be regarded as a trend, due to a need of more refined meshes for this value of K_{G}.

The parameter K_{G} has been varied within six orders of magnitude. This corresponds to three orders of magnitude of the internal length scale parameter $l = \sqrt{K_{\mathrm{G}}/E}$. Considering this range of change, one might have expected a stronger effect of the variation of K_{G} on the overall material behavior. However, it should be noted that the considered load case (tension) corresponds to a minimum of plastic heterogeneity.

The tension and torsion simulations of Wulfinghoff and Böhlke (2012a) indicate that more heterogeneous deformation modes (like torsion) imply a stronger effect of K_G on the mechanical response, mainly on the overall hardening (a detailed consideration exceeds the scope of the work at hand, for details see Wulfinghoff and Böhlke (2012a)). As a result, the overall tensile response is only weakly affected by K_G, but mainly controlled by the grain boundary yield strength Ξ_0^C.
The parameters K_G and Ξ_0^C control the size dependence of the model. In order to illustrate this size effect, Fig. 5.5 shows the influence of the cube length L on the mechanical response. For decreasing size the model responds with increased strength. Upon letting the gradient contribution to the stored energy as well as the grain boundary yield strength vanish, it can be observed that this size effect disappears (see the lower three coinciding curves).
With the aforementioned interpretations of the material parameters at hand, the material model can be fitted to experimental tensile test data based on the following guidelines.

- The size dependence of the overall yield stress is mainly controlled by the grain boundary yield strength Ξ_0^C. Therefore, Ξ_0^C must be fitted to the vertical distances of the stress strain curves associated to different specimen and/or grain dimensions in the plastic regime.
- For sufficiently large values of the strain rate sensitivity p, the material model approximates a rate-independent model. In this case, a moderate variation of the reference strain rate $\dot{\gamma}_0$ has no noticeable effect. Additionally, the offset stress (i.e. the yield stress of the lower three (coinciding) curves in Fig. 5.5) is mainly controlled by $\tau_0^C + \tau^D$ in the rate-independent limit.
- The Voce hardening parameters Θ_0 and τ_∞^C control the evolution of the hardening slope.
- As has been discussed, a meaningful fitting of the parameter K_G requires experimental data of more heterogeneous deformation modes than tension.

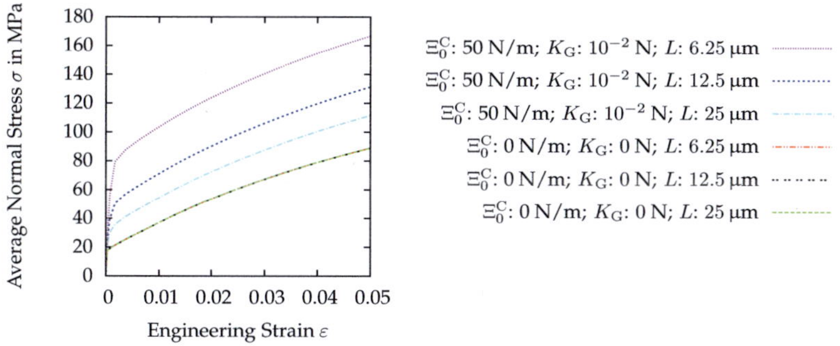

Figure 5.5: Influence of the structure size on the mechanical response of an oligocrystal with 8 grains in the presence and absence of grain boundary and gradient influence under tensile load.

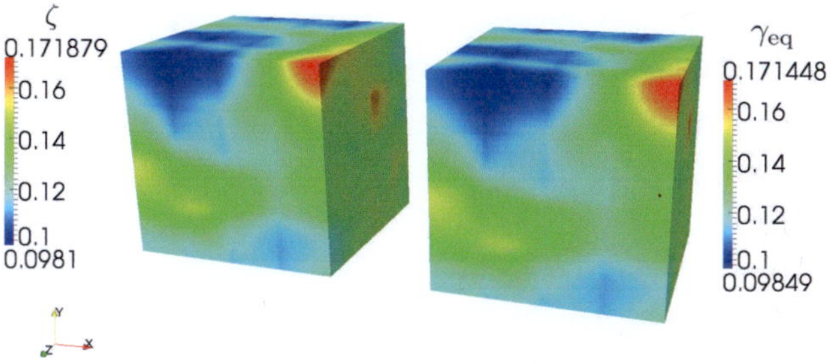

Figure 5.6: Resulting distributions of the equivalent plastic strain γ_{eq} and its micromorphic counterpart ζ for a tensile test simulation with 64 grains.

In order to illustrate the influence of the penalty approximation controlled by H_χ, the distributions of ζ and γ_{eq} are depicted in Fig. 5.6 for one of the simplified grain aggregate simulations. Although small local deviations occur for the selected penalty parameter, the overall field solutions of both quantities are still in reasonable agreement. In addition, in Tab. 5.3 the influence of the presence of the interface yield criterion on the computational costs is shown. There is no obvious trend, regarding the influence of the grain boundary yield strength Ξ_0^C on the computational time.

Table 5.3: Computational time of a 64-crystal tensile test simulation for different values of the parameter Ξ_0^C, where "0" corresponds to a simulation without grain boundary resistance. Calculations are performed on Intel Xeon CPU E5540 2.53GHz.

Ξ_0^C	Number of timesteps	Largest timestep	Computational time	DOFs
0 N/m	12	0.32 s	18042 s	275684
50 N/m	10	0.35 s	17112 s	275684
100 N/m	15	0.16 s	28786 s	275684

5.3.4 Comparison of Simulation Results to Experimental Data

In the following a work by Yang et al. (2012) is considered, focusing on the size effects of polycrystalline copper microwires under tensile load. The mechanical response of three microwires annealed at different temperatures (and thus having substantially different average grain sizes) is simulated by simplified cubic grain aggregates, cf. Fig. 5.2. The grain sizes are chosen in a way, such that the average grain size reported in the experiments is approximately matched in the first two cases, cf. also Fig. 5.7. Thus, the microstructures are accounted for in a simplified, yet reasonable way. In the third case, one grain with a length of 20 μm is chosen for simplicity and thus the grain size is within the tolerance of the reported average grain size. Consequently, this leads to an additional smaller grain of 5 μm in the setup.
The mechanical response is fitted (with the described guidelines in the preceding subsection) to the experimental data of the annealed 25 μm thick microwires (see Figure 3 (a) in Yang et al. (2012)), and the resulting simulated stress-strain curves are shown in comparison in Fig. 5.7. Furthermore, the distributions of the micromorphic approximation of the equivalent plastic strain at an engineering strain of 0.05 are depicted for all three cases in Fig. 5.8. The determined plastic model parameters for the fitted tensile curves as well as the used elastic material

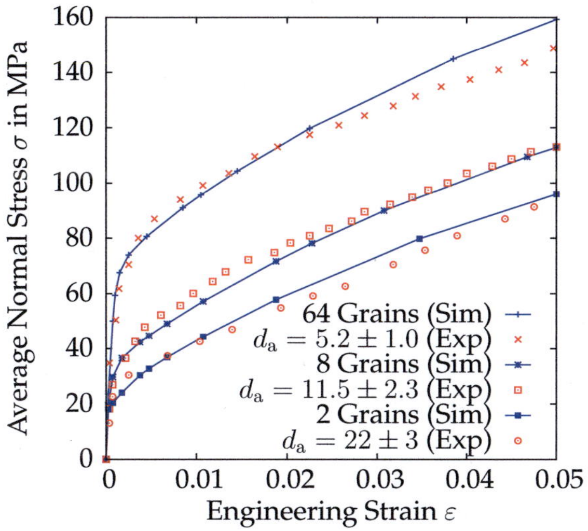

Figure 5.7: Comparison of simulation results of simplified grain aggregates (Sim) and experimental data (Exp) of copper microwire tensile tests by Yang et al. (2012). From the experiments of Yang et al. (2012), the average grain size - here denoted by d_a in μm - (which was there determined as half of the mean of the longitudinal and transversal average grain size) is listed above.

parameters for copper are listed in Tab. 5.2. Assuming a macroscopic Young's modulus of $E = 100\,\mathrm{GPa}$, the internal length scale parameter is given by $l = \sqrt{K_G/E} \approx 0.32\mu\mathrm{m}$. This value of K_G has the same order of magnitude as in Wulfinghoff and Böhlke (2012a), where no grain boundary yield condition was accounted for. As has been discussed in the previous section, the parameter K_G can be varied in a wide range without having a strong effect on the results since a tensile test is considered. A more rigorous determination of K_G requires experimental data of more heterogeneous deformations.

In addition, in Fig. 5.9, for an 8 grain aggregate the distributions of the equivalent plastic strain and the gradient hardening stress $\mathrm{div}(\boldsymbol{\xi})$ are depicted in comparison for the first four time increments. As soon as the grain boundary yield criterion is fulfilled, the grain boundaries start

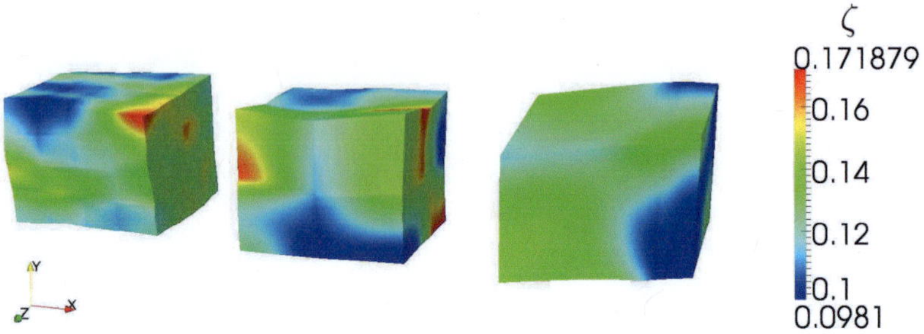

Figure 5.8: Resulting distributions of the micromorphic approximation of the equivalent plastic strain for tensile test simulations of simplified grain aggregates; displacement scaled by factor 3.

yielding. It can be seen in Fig. 5.9 that during each time increment the amount of plastically activated grain boundary nodes is increased.

5.3.5 Discussion of Results

The stress-strain curves of the simulated grain aggregates are in good agreement with the experimental data, cf. Fig. 5.7. It is emphasized that for all three simulations the same set of parameters is used and consequently the model represents the experimentally observed size effects qualitatively and quantitatively.

As can be seen here, for the presented set of parameters, there exists a correlation between the grain boundary area in a unit volume and the initial yielding resistance of the sample. The more grain boundaries, or rather the more grain boundary area is present, the higher is the initial resistance to yielding.

The depicted gradient hardening stress distributions in Fig. 5.9 show an increase close to the grain boundaries with proceeding plastic deformation. However, it should be noted that the rate of increase of the gradient hardening stress $\mathrm{div}\,(\boldsymbol{\xi})$ is higher in the beginning of the deformation process than in the end.

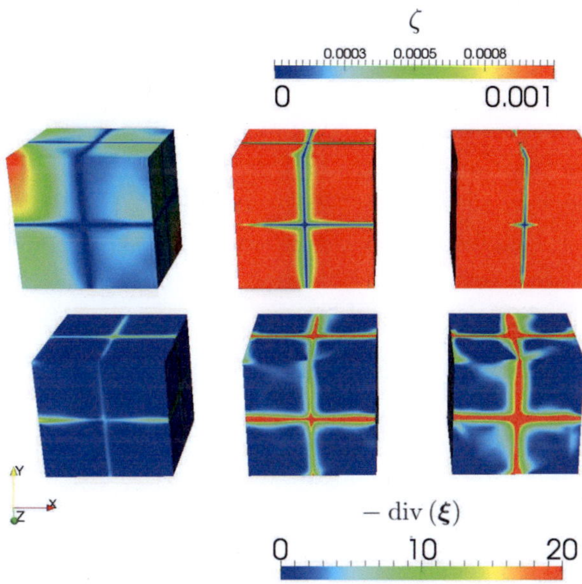

Figure 5.9: Distributions of the micromorphic approximation of the equivalent plastic strain γ_{eq} and the gradient hardening stress div $(\boldsymbol{\xi})$ for tensile test simulations of a simplified grain aggregate with 8 grains. Depicted are the distributions at the end of each of the first four timesteps. The color scale of the upper series is limited in order to visualize the grain boundary yielding.

5.4 Enhanced Power-Law Subroutine

5.4.1 Convergence Improvement

This section deals with the theory developed in Sects. 5.1 to 5.2 and focuses on the implementation details of the associated power-law material subroutine. It is to a large part taken from Wulfinghoff and Böhlke (2013). For applications of the theory to periodic polycrystals, see Wulfinghoff et al. (2013b,c).

It turns out that the gradient enhancement supports the improvement of the convergence properties of the power law subroutine. As a result of the gradient extension, a reasonable guess of the starting solution for

the local Newton iteration scheme can be identified. This allows for significantly larger time steps.
The local material subroutine always converged in the subsequently illustrated simulations, even in the case of strongly nonlinear material properties, large time steps and two million degrees of freedom. An exemplary performance comparison of the material subroutine with and without improved initial solution guess demonstrates the computational benefit of the approach.

5.4.2 Improvement of the Starting Solution

The numerical solution of the nonlinear system of equations (5.24) and (5.25) is challenging in the case of large values of the rate sensitivity parameter p. The evaluation of the power term is not always possible since, in practice, the result often exceeds the numerical range of a standard computer. This problem can occur if the initial guess for the local Newton scheme is not sufficiently close to the correct solution. If the time increment Δt is small, the solution of the last step can be expected to be a good starting value. However, increasing the time step rapidly leads to a loss of convergence.
The algorithm can be stabilized by improving the starting solution as follows: in a first step the flow rule (5.18) is modified in the large stress range by a linear function (cf. Fig. 5.10). A continuous differentiable transition between power law and linear approximation is used at the point indicated by the parameter $\dot{\gamma}_{\mathrm{L}}$. Thereby, regularized versions of Eqns. (5.24) and (5.25) are obtained which are less sensitive concerning the numerical range of the computer. The solution of the regularized problem can be expected to be an improved starting solution for the correct (i.e. non-regularized) system of equations (5.24) and (5.25).
The smaller the parameter $\dot{\gamma}_{\mathrm{L}}$ is chosen the smaller is the slope of the linear approximation and, consequently, the better is the convergence of the regularized model. In contrast, $\dot{\gamma}_{\mathrm{L}}$ should not be chosen too small since the regularized model (and improved starting value) is supposed

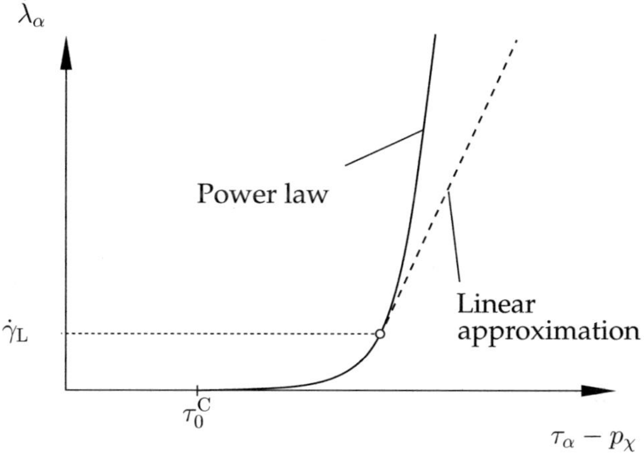

Figure 5.10: Regularization of the power law for the computation of an improved starting solution for the Newton scheme.

to be as close as possible to the correct model and the associated solution. To satisfy both requirements at once the value of $\dot{\gamma}_L$ is estimated by

$$\dot{\gamma}_L = \frac{k_L(|\Delta\zeta| + \epsilon)}{\Delta t}, \tag{5.35}$$

where $k_L > 1$ is a factor close to 1 and ϵ is a small number. In this work, the values $k_L = 2$ and $\epsilon = 10^{-4}$ are applied. This choice permits the solution of the regularized problem to be (in most cases) within the power law range, since for large values of the penalty parameter H_χ, the following relation holds

$$\frac{\Delta\lambda_\alpha}{\Delta t} \leq \frac{\Delta\gamma_{eq}}{\Delta t} \approx \frac{\Delta\zeta}{\Delta t} \leq \dot{\gamma}_L. \tag{5.36}$$

At the same time, the definition (5.35) of the parameter $\dot{\gamma}_L$ is expected to be sufficiently small to prevent the algorithm from exceeding the numerical range of the computer.

The choice of the numerator $k_L(|\Delta\zeta| + \epsilon)$ in estimation (5.35), instead of simply using $\Delta\zeta$, is motivated by the fact that the approximation $\Delta\gamma_{eq} \approx \Delta\zeta$ in Eq. (5.36) is not exact.

Having obtained the improved starting guess, the correct solution is obtained in a subsequent Newton iteration by increasing $\dot{\gamma}_{\mathrm{L}}$ to a very large value to assure that the solution lies in the power law (not in the linear) regime. Indeed, the numerical examples discussed in the next section indicate significantly improved convergence properties.

5.4.3 Numerical Examples

Simulation Set-Up

The performance of the improved power law subroutine is compared to the standard subroutine in a tensile test simulation. Periodic boundary conditions are applied, i.e. the displacement field is assumed to be given by

$$\boldsymbol{u} = \bar{\boldsymbol{\varepsilon}}\boldsymbol{x} + \tilde{\boldsymbol{w}}, \tag{5.37}$$

where $\bar{\varepsilon}$ represents the macroscopic strain and $\tilde{\boldsymbol{w}}$ the displacement fluctuations. The variable ζ is assumed periodic. Applying a tensile test with free lateral contractions, the shear strains $\bar{\varepsilon}_{ij},\ i \neq j$ are set to zero. The tensile strain $\bar{\varepsilon}_{11} = \varepsilon(t)$ is prescribed and $\bar{\varepsilon}_{22}$ as well as $\bar{\varepsilon}_{33}$ represent additional degrees of freedom.
In order to model the gradient and isotropic hardening contributions, the quadratic defect energy $W_{\mathrm{g}}(\nabla\zeta) = 1/2\, K_{\mathrm{G}} \nabla\zeta \cdot \nabla\zeta$ and the Voce hardening relation (5.34) are applied. The plastic material parameters are given by

$\dot{\gamma}_0$	p	τ^{D}	τ_0^{C}	τ_∞^{C}	Θ_0	K_{G}	H_χ
$10^{-3}\mathrm{s}^{-1}$	10	1 MPa	70 MPa	200 MPa	1 GPa	0.01N	10^7MPa

Here, τ_∞^{C} is the limit yield stress and Θ_0 is an initial hardening modulus. The elastic constants read $C_{1111} = 168\,\mathrm{GPa}$, $C_{1122} = 121\,\mathrm{GPa}$ and $C_{1212} = 75\,\mathrm{GPa}$, respectively. To prevent the standard material subroutine (i.e. the non-improved algorithm) from requiring excessively small time steps, the rate sensitivity exponent is chosen relatively small ($p = 10$). The tolerances of the global Newton scheme and the material subroutine are given by 10^{-9} (times the initial euclidean

residual norm) and 10^{-10} (absolute), respectively. However, numerical experiments indicate that much (several orders of magnitude) less demanding tolerances already yield satisfactory results which can barely be distinguished from the presented findings. If no convergence can be achieved at the global or at the integration point level, the time step is reduced, otherwise it is increased (by a factor of ~ 2).
Two periodic FCC grains (see Fig. 5.11) with random orientations are discretized by $32 \times 32 \times 32$ standard linear hexahedrons with ~ 130000 degrees of freedom in total. The cube shaped domain has an edge length of $30\ \mu m$. In order to apply a simple geometric multigrid solver, a penalty approximation of the micro-hard boundary conditions was implemented, i.e., the boundary condition $\zeta = 0$ is approximated. The applied strain rate is $0.05\,s^{-1}$.

Discussion of the Results

The macroscopic tensile stress response is visualized in Fig. 5.12. The simulation results of the standard and the improved integration point subroutine match qualitatively and quantitatively, i.e., there is neither a dependence on the algorithm nor on the time step size observable. This is underlined by a comparison of the final deformations of both simulations in Fig. 5.11.

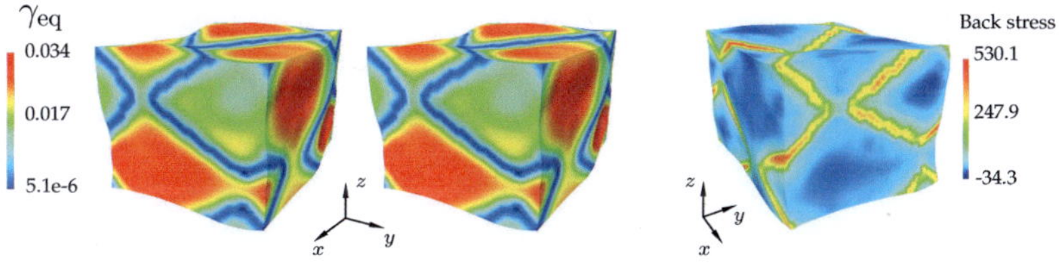

Figure 5.11: Comparison of the simulation results of the standard (left) and improved (center) algorithm. Right: visualization of the back stress $-\mathrm{div}\,(\boldsymbol{\xi})$.

Fig. 5.12 shows that the standard algorithm requires significantly smaller time steps than the improved Newton scheme. Especially, the numerically challenging elasto-plastic transition diminishes the time

step size of the standard scheme due to loss of convergence of the integration point Newton procedure. Contrary, no convergence problems are induced at the transition in case of the improved scheme. In total, the standard algorithm required 158 time steps, while the improved scheme always converged leading to a total number of five steps. The convergence properties of the non-improved algorithm can be ameliorated by decreasing the strain rate sensitivity p, as illustrated in Fig. 5.12 (right). Consequently, larger deformations (here 5% macroscopic strain) can be simulated in passable times, even with the standard routine.
Table 5.4 (left) summarizes the convergence of the improved scheme for $p = 10$. The convergence of the Newton scheme can be improved by increasing the accuracy of the linear equation solver (by increasing the maximum number of V-Cycles). Table 5.4 (right) shows exemplary convergence rates for this case. However, since the solution of the individual linear systems is more time consuming, a less accurate solution of the multigrid solver is accepted leading to an increased number of global Newton steps.

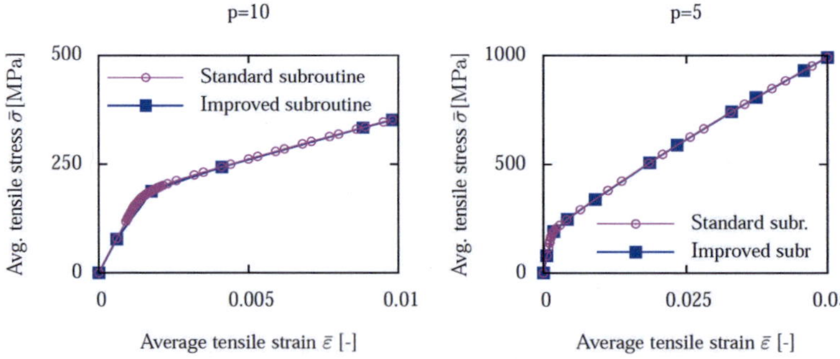

Figure 5.12: Left: Macroscopic tensile response computed by the standard and the improved algorithm for $p = 10$. Right: Amelioration of the convergence properties of the non-improved algorithm for a decreased strain rate sensitivity $p = 5$.

In the following, a physical interpretation of the results is summarized.

- The back stress plays an important role close to the grain bound-

Step 1	Step 2	Step 3	Step 4	Step 5
1.00e+00	1.00e+00	1.00e+00	1.00e+00	1.00e+00
3.95e-03	2.79e-02	1.56e-02	2.76e-02	2.95e-02
1.89e-05	1.01e-02	8.15e-03	1.27e-02	6.75e-03
2.80e-07	2.97e-02	3.08e-03	2.35e-03	1.50e-03
4.18e-09	4.52e-04	2.03e-03	5.53e-04	4.44e-04
6.36e-11	1.84e-04	1.16e-03	1.33e-04	6.92e-05
	1.04e-04	3.82e-04	1.14e-05	5.00e-06
	4.10e-05	5.20e-05	5.03e-07	2.62e-07
	1.44e-05	1.36e-06	9.31e-09	2.09e-09
	8.16e-06	6.71e-09	8.47e-11	1.72e-11
	7.42e-06	6.11e-11		
	1.72e-07			
	1.14e-09			
	1.70e-11			

Step 1	...	Step 5
1.00e+00		1.00e+00
4.33e-07		6.68e-04
2.40e-13		3.86e-05
		2.57e-06
		3.34e-08
		8.98e-12

Table 5.4: Left: Euclidean norm (normalized) of the residual of the improved algorithm. The values belong to the results in Fig. 5.12. Right: exemplary convergence rates illustrating that increasing the accuracy of the linear equation solver tends to decrease the number of necessary Newton steps (Steps 2, 3 and 4 required 14, 10 and 7 Newton iterations, respectively).

aries (Fig. 5.11), where it significantly reduces the plastic deformation. Physically, this result represents the strong dislocation interaction forces in dislocation pile-ups. This behavior is typical of strain gradient plasticity theories.

- The macroscopic stress-strain response (Fig. 5.12) is comparable to the prediction of other gradient plasticity theories. Especially, the quadratic defect energy leads to an additional, approximately linear hardening contribution (for a detailed discussion cf. Wulfinghoff and Böhlke (2012a) and Sect. 1.5.3).

Towards Polycrystal Simulations

In the following, the performance of the improved material subroutine is tested in the case of a large strain rate sensitivity exponent ($p = 200$) as well as an elevated number of grains (with random orientations) and elements.

Additionally, the ultimate tensile strain is increased to $\bar{\varepsilon}_{11}(1\mathrm{s}) = 0.05$. The edge length of the cube shaped domain is $60\ \mu\mathrm{m}$. Figs. 5.13 and 5.14 illustrate the results of a periodic micro structure consisting of 27 grains which are discretized by $80 \times 80 \times 80$ elements. Consequently, the total number of degrees of freedom is $\sim$2 million. Except for these

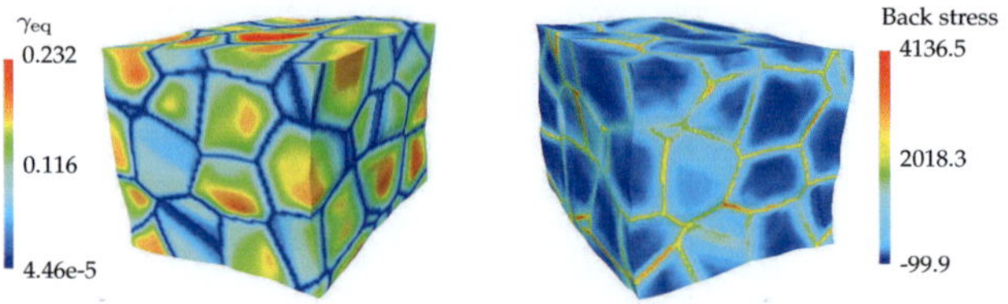

Figure 5.13: Deformed periodic micro structure at 5% macroscopic tensile strain.

changes, all model parameters remain unchanged. The total number of integration points is ~4 million. The associated material subroutines converged in all time steps and all iterations.

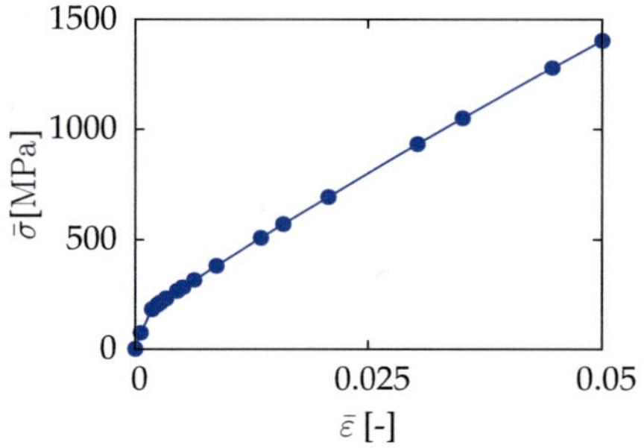

Figure 5.14: Macroscopic tensile response of the model in Fig. 5.13.

However, the global Newton scheme did not always converge leading to a reduction of the time step size, especially in the elasto-plastic transition phase (cf. Fig. 5.14). The simulation was done on a Pentium Dual Core PC with 3.0GHz and 6 GB RAM. Neither the multigrid solver, nor the integration point subroutine evaluation was parallelized. The global and local convergence tolerances of 10^{-9} and 10^{-10} (see above) were not changed. The total simulation time was ~41 hours.

Chapter 6

Gradient Plasticity Grain Boundary Models vs. a Classical Pile-Up Theory

6.1 Introduction

The Hall-Petch effect (Hall, 1951; Petch, 1953) is a central motivation of gradient plasticity theories. Many frameworks include a plastic grain boundary energy, see e. g. Fredriksson and Gudmundson (2005), Aifantis and Willis (2005) or Aifantis et al. (2006). Other works using interface energies in the context of gradient plasticity are by Fredriksson and Gudmundson (2007) and Voyiadjis and Deliktas (2009). Gurtin (2008b) suggests a theory that accounts automatically for grain misorientations. The theory developed by Dahlberg et al. (2013) models interfaces that slide and separate.

The present chapter aims at an investigation concerning the potential of certain gradient plasticity interface models to mimic the Hall-Petch effect. It is found that the macroscopic yield stress and the grain size of these models are indeed inversely related. However, in general, the increase of the yield stress of the gradient models does not scale like $1/\sqrt{D}$ (where D is the mean grain diameter).

The origin of this apparently wrong scaling behavior of the models is investigated. Therefore, the gradient theories are compared to the work of Eshelby et al. (1951) (see also Armstrong et al. (1962)). Their work represents one of the most prominent explanations of the Hall-Petch effect. In addition, the gradient models and the Eshelby model account for similar phenomena, like pile-ups and grain boundary yielding. Therefore, it seems to be worthwile to compare both approaches. Moreover, the work at hand is driven by the desire to improve the understanding of gradient plasticity.

The outline of the chapter is as follows. First, the scaling behavior of certain grain boundary models discussed. Subsequently, the work of Eshelby et al. (1951) is reviewed. Their work is then compared to certain grain boundary models in gradient plasticity. In particular, the grain boundary stresses of both models are compared, since these govern the grain boundary yielding behavior. The stresses are found to be related, but in general inconsistent. This leads to different Hall-Petch exponents, i.e., different scaling laws.

The findings are subsequently applied to a gradient plasticity grain boundary yield theory, which is close to the model formulated in Chap. 5. Three dimensional polycrystal simulations with several hundred grains illustrate and confirm the theoretical results.

6.2 Size Effects provoked by Energetical and Dissipative Grain Boundary Models within Gradient Plasticity

The size effects due to energetical and dissipative grain boundary models in gradient plasticity are investigated in the following. Energetical grain boundaries are assumed to store energy as a result of plastic deformations at the grain boundary. In contrast, the power generated at dissipative grain boundaries is not stored. The grain size dependence of the macroscopic yield stress σ_Y of the gradient models is of particular interest.

For real materials, this dependence is well approximated for many metals by the Hall-Petch relation (Hall, 1951; Petch, 1953)

$$\sigma_\text{Y} = \sigma_\text{Y0} + \frac{k_\text{HP}}{\sqrt{D}}, \tag{6.1}$$

where σ_Y0 denotes the yield stress of a very coarse-grained polycrystal, k_HP is the Hall-Petch slope and D is the average grain diameter.

The Hall-Petch relation is the most common relation describing grain boundary strengthening. However, it should be mentioned, that doubts concerning its generality exist, in particular with respect to the exponent of -0.5 (e.g. Dunstan and Bushby, 2014).

6.2.1 Energetical Interfaces

Despite the formal simplicity of the Hall-Petch relation (6.1), the formulation of associated gradient plasticity models turns out to be non-trivial. Interface energy terms have been introduced into the theory to model the grain boundary (GB) resistance, e.g., in Chaps. 4 and 5. In the following, it is investigated if such energies are consistent with the Hall-Petch relation (6.1).

A continuum model formulation on the microscale is considered, i.e., individual grains or phases, but no discrete dislocations are resolved. The GB energy density is assumed to be a convex function of some plastic strain variable ε^{p}

$$W_\Gamma = W_\Gamma(\varepsilon^{\mathrm{p}}), \quad W_{\Gamma,\mathrm{tot}} = \int_\Gamma W_\Gamma \, \mathrm{d}a. \tag{6.2}$$

For simplicity, ε^{p} is assumed to be a scalar variable (this might be the equivalent plastic strain, for example, see Chaps. 4 and 5). However, the considerations are expected to be formally similar for tensorial variables. The set Γ denotes the union of all grain boundaries. The rate of W_Γ reads

$$\dot{W}_\Gamma = \Xi_\Gamma^{\mathrm{e}} \, \dot{\varepsilon}^{\mathrm{p}}, \tag{6.3}$$

where $\Xi_\Gamma^{\mathrm{e}} = \partial_{\varepsilon^{\mathrm{p}}} W_\Gamma$ is an energetic force.

The introduction of W_Γ changes the macroscopic/overall response of the polycrystal model. In the region of well-established flow, it is expected to increase the macroscopic yield stress. The following scaling behavior is estimated (see App. C.1)

$$\Delta\bar{\sigma} \sim \frac{1}{D}. \tag{6.4}$$

This result contrasts with the Hall-Petch relation (6.1), where the scaling behavior is $1/\sqrt{D}$. Note that the form of W_Γ (for example homogeneous of degree one or quadratic in $\varepsilon^{\rm p}$) does not influence the result (6.4).

6.2.2 Dissipative Interfaces

Dissipative interfaces are treated in analogy with energetic interfaces. The dissipation per unit grain boundary area is assumed to be given by

$$\mathcal{D}_\Gamma = \Xi_\Gamma^{\rm d}\,\dot{\varepsilon}^{\rm p}. \tag{6.5}$$

Here, $\Xi_\Gamma^{\rm d}$ is a dissipative force.
In App. C.2, it is assumed that $\Xi_\Gamma^{\rm d}$ is constant in the region of well-established plastic flow. In this case, the resulting scaling behavior is approximately given by (App. C.2)

$$\Delta\bar{\sigma} \sim \frac{1}{D}. \tag{6.6}$$

Again, the result contrasts with the Hall-Petch relation (6.1).
As a summary, neither the energetical grain boundaries nor the proposed dissipative model lead to the scaling behavior $1/\sqrt{D}$. The work at hand tries to investigate why these models exhibit a different scaling behavior.

6.3 Review of the Hall-Petch Model of Eshelby et al. (1951)

If a polycrystal is loaded by external forces, the body initially deforms mainly elastically. When the internal stresses become sufficiently large, a significant number of dislocations starts to move inside of the crystals. It might be assumed that initially, most grain boundaries represent hard obstacles for dislocations. This is usually mainly due to the lattice misfit. When the external loading is increased, the dislocation driving forces become larger. At some point, dislocations are assumed to enter, leave and interact with the grain boundary. Subsequently, this process is called grain boundary yielding. In the following, an associated yield criterion will be formulated. Therefore, an arbitrary grain is considered and modeled as a sphere with diameter d. It is assumed that a slip band is formed in the center of the grain during the initial loading period. For simplicity, one slip system is considered. The area of the circular slip band is given by $A = \pi d^2/4$.

The dislocations in the slip band are assumed to build pile-ups at the grain boundary. This kind of dislocation configuration induces stress concentrations in the vicinity of the grain boundary. According to Eshelby et al. (1951), these stresses are comparable to the stress field of a crack tip. Nearby dislocation sources experience this increased stress intensity. Such a nearby source in the adjacent grain is considered. If it is appropriately oriented, it can be activated. This means, it emits dislocations and thereby extends the slip band into the next grain.

This propagation of the slip band from grain to grain is considered as the starting point of increased plastic deformation (see, e.g., Armstrong et al. (1962)). It is further assumed that, on the macro scale, this propagation mechanism causes the elasto-plastic transition in the stress-strain diagram.

In the following, the initial activation of the aforementioned source is considered. The source is idealized as a point. It is assumed to be activated if the absolute value of the resolved shear stress $\tau_{\rm s}$ at the source reaches a critical value $\tau_{\rm s}^{\rm C}$. Therefore, the elastic range of the source is given by

$$\tau_{\rm s} \leq \tau_{\rm s}^{\rm C}. \tag{6.7}$$

The grain boundary yield point is reached when the source starts to emit dislocations. At that point, $\tau_{\rm s}$ reaches the critical value $\tau_{\rm s}^{\rm C}$ and the plastic deformation is transmitted through the grain boundary. Therefore, the grain boundary yield criterion can be formulated as

$$f_\Gamma^* = \frac{1}{4k_1}(|\tau_{\rm s}| - \tau_{\rm s}^{\rm C}) \leq 0. \tag{6.8}$$

The constant factor $1/(4k_1)$ will be explained subsequently. Eshelby et al. (1951) found a relation close to the following one

$$|\tau_{\rm s}| = k_1 |\bar{\tau}_{\rm eff}| \sqrt{d}. \tag{6.9}$$

Here, k_1 is a constant of proportionality, which depends on the distance of the source from the end of the pile-up. The average effective resolved shear stress $\bar{\tau}_{\rm eff}$ in the slip band is given by

$$\bar{\tau}_{\rm eff} = \frac{1}{A} \int_A (\tau - \tau^{\rm d})\, {\rm d}a. \tag{6.10}$$

Here, $\tau^{\rm d}$ models dissipative friction-like effects and τ denotes the resolved shear stress (the projection of the Cauchy stress onto the slip system). Note that the work of Eshelby et al. (1951) is not based on the average effective shear stress $\bar{\tau}_{\rm eff}$. Instead, the more simple (one-dimensional) setting of a homogeneous stress is assumed (which, however, coincides with the average stress in that case).
For simplicity, friction-like effects are neglected in the following

$$\tau^{\rm d} = 0 \Rightarrow \bar{\tau}_{\rm eff} = \bar{\tau} = \frac{1}{A} \int_A \tau \, {\rm d}a. \tag{6.11}$$

In this case, the combination of Eqns. (6.7) and (6.9) gives the following grain boundary yield condition

$$|\bar{\tau}| = \frac{|\tau_{\rm s}|}{k_1 \sqrt{d}} \le \frac{\tau_{\rm s}^{\rm C}}{k_1 \sqrt{d}}. \tag{6.12}$$

If it is assumed that $\bar{\tau}$ is proportional to the externally applied load, Eq. (6.12) directly yields the Hall-Petch relation. Note that Eq. (6.12) has been derived for a central slip band. It can be argued that a similar result also holds for non-central slip bands. In this case, k_1 would take a different value.
In the following, the aforementioned model will be referred to as Eshelby model, since it is closely related to the work of Eshelby et al. (1951).

6.4 Comparison of the Eshelby Model with Gradient Plasticity Grain Boundary Models

As pointed out in Sect. 6.2, the considered gradient plasticity interface models are expected to imply the scaling behavior $\Delta\bar{\sigma} \sim 1/D$ (Eqns. (6.4) and (6.6)). In contrast, the Eshelby model implies $\Delta\bar{\sigma} \sim 1/\sqrt{D}$. In order to compare the gradient models with the Eshelby model, the former are reviewed in the following.

6.4.1 Review of Grain Boundary Stresses in Gradient Plasticity

In many gradient plasticity theories, the number of balance equations is extended. Typically, micro force balance equations arise in addition to the linear momentum balance. Here, it is assumed that this additional balance equation is given by (see, e.g., Gurtin et al. (2007) and Sect. 1.5, Eq. (1.27))

$$\tau + \operatorname{div}\left(\tilde{\boldsymbol{\xi}}\right) - \tau^{\mathrm{d}} = 0, \tag{6.13}$$

where $\tilde{\boldsymbol{\xi}}$ is a so-called microstress. Here, this quantity is assumed to depend on the gradient of the plastic shear strain $\nabla\gamma$ projected onto the slip plane. Some quantities are marked by $(\tilde{\bullet})$ in order to distinguish them from similar quantities introduced in subsequent sections.

In this section, the out-of-(slip-)plane component of $\tilde{\boldsymbol{\xi}}$ is assumed to vanish (like in most approaches). The presence of microstresses implies the existence of microtractions

$$\tilde{\Xi} = \tilde{\boldsymbol{\xi}} \cdot \boldsymbol{n} \tag{6.14}$$

playing an important role at boundaries. Here, $\boldsymbol{n}$ denotes the outer surface normal.

The solution of strain gradient plasticity problems of the aforementioned type necessitates the definition of boundary conditions. Very often, γ is simply set to zero in order to model boundaries impenetrable for dislocations. In this case, a steady increase of $|\tilde{\Xi}|$ at the GB with ongoing deformation can often be observed, especially if the strain gradients at the grain boundary take large values. Physically, the increased strain gradients normally represent dislocation pile-ups. The generalized force $|\tilde{\Xi}| = |\tilde{\boldsymbol{\xi}} \cdot \boldsymbol{n}|$ is the larger the more dense the dislocations at the grain boundary, since $\tilde{\boldsymbol{\xi}} = \tilde{\boldsymbol{\xi}}(\nabla\gamma)$ usually increases with increasing gradients.

As a consequence, it seems to be natural to assume that a critical value of $|\tilde{\Xi}|$ leads to grain boundary yielding, i.e. $|\tilde{\Xi}|$ is interpreted as grain boundary load (see also the discussion in Sect. 3.6.2).

It should be noted that, in general, microtractions $\tilde{\Xi}$ of both grains and all slip systems must be considered. However, this is not necessary if it is assumed that the slip system in the considered grain is the only active one, yet.

A more formal interpretation of $|\tilde{\Xi}|$ as grain boundary loading may be based on the requirement of positive dissipation, i.e., a grain boundary yield criterion with $|\tilde{\Xi}|$ as grain boundary loading is expected to yield a thermodynamically consistent theory. This means that the microtraction $|\tilde{\Xi}|$ represents the driving force of plastic grain boundary deformation processes. Plastic deformations are assumed to occur when the associated driving forces reach a critical value. In order to further clarify this argumentation, it will be applied to a specific gradient model in Sect. 6.5.

6.4.2 Comparison with the Eshelby Model

In the following, a comparison of the gradient and the Eshelby model is proposed. In particular, the grain boundary loading stresses of both theories ($|\tau_{\mathrm{s}}|$ and $|\tilde{\Xi}|$) are compared. This can be achieved based on the observation that the average effective stress $\bar{\tau}_{\mathrm{eff}}$ can be expressed in terms of the microtractions. From Eqns. (6.10) and (6.13) it follows that

$$\bar{\tau}_{\mathrm{eff}} = \frac{1}{A} \int_A (\tau - \tau^{\mathrm{d}}) \, \mathrm{d}a \overset{(6.13)}{=} -\frac{1}{A} \int_A \mathrm{div}\left(\tilde{\boldsymbol{\xi}}\right) \, \mathrm{d}a. \qquad (6.15)$$

Since the component of $\tilde{\boldsymbol{\xi}}$ which is normal to the glide plane is assumed to vanish, Gauss' theorem can be applied

$$\bar{\tau}_{\mathrm{eff}} = -\frac{1}{A}\int_{\partial A} \tilde{\boldsymbol{\xi}} \cdot \boldsymbol{n}\, \mathrm{d}s = -\frac{1}{A}\int_{\partial A} \tilde{\Xi}\, \mathrm{d}s. \tag{6.16}$$

For the special case of constant microtractions at the grain boundary, it follows that

$$\bar{\tau}_{\mathrm{eff}} = \frac{-\pi d\, \tilde{\Xi}}{\pi (d/2)^2} = -\frac{4\, \tilde{\Xi}}{d}, \tag{6.17}$$

since the grain is assumed spherical.

In order to compare the grain boundary loading stresses of both models, Eqns. (6.8), (6.9) and (6.17) are combined, yielding the following result

$$f_{\Gamma}^{*} = \frac{1}{4k_1}(|\tau_{\mathrm{s}}| - \tau_{\mathrm{s}}^{\mathrm{C}}) \leq 0, \qquad |\tau_{\mathrm{s}}| = \frac{4k_1}{\sqrt{d}}\, |\tilde{\Xi}|. \tag{6.18}$$

Eq. (6.18) is the central result of this chapter. It is interpreted as follows: the quantity $|\tau_{\mathrm{s}}|$ describes the stresses at dislocation sources in the vicinity of the grain boundary (in the Eshelby model). These stresses represent the grain boundary loading, since the sources emit dislocations at the point where $|\tau_{\mathrm{s}}|$ reaches a critical value $\tau_{\mathrm{s}}^{\mathrm{C}}$. This point is considered as the grain boundary yield point. Mathematically, it is described by the yield condition f_{Γ}^{*}.

Equation $(6.18)_2$ relates the grain boundary loading $|\tau_{\mathrm{s}}|$ of the Eshelby model with the grain boundary loading $|\tilde{\Xi}|$ of the gradient theory. These two GB stress quantities are found to be proportional (in the simple setting considered). Therefore, the interpretation of $|\tilde{\Xi}|$ as a grain boundary load is closely connected to the respective interpretation of $|\tau_{\mathrm{s}}|$. In particular, a gradient plasticity simulation of a given system might allow to roughly estimate the stresses $|\tau_{\mathrm{s}}|$ of a respective system in the spirit of Eshelby et al. (1951).

However, according to Eq. (6.18), $|\tau_{\mathrm{s}}|$ is not only proportional to $|\tilde{\Xi}|$ but also to $1/\sqrt{d}$. This result suggests that the grain boundary loading terms ($|\tau_{\mathrm{s}}|$ on the one hand and $|\tilde{\Xi}|$ on the other) of the two theories are not fully compatible. The microtractions $|\tilde{\Xi}|$ cannot be considered as a universal measure of the source stresses $|\tau_{\rm s}|$, since the term $\sqrt{d}$ has to be considered, too. If this term is not accounted for, the scaling behavior of the gradient model will be, in general, different from the scaling behavior $1/\sqrt{d}$ of the Eshelby model (as outlined in Sect. 6.2). Of course, the explicit consideration of a characteristic microstructural length like the grain diameter is not desirable if a gradient plasticity theory is applied.

As a result, a gradient plasticity simulation of a *specific* microstructure might be used to roughly estimate the GB loading $|\tau_{\rm s}|$ of an associated Eshelby model. This can be done by evaluating the computed GB microtractions $|\tilde{\Xi}|$ in combination with Eq. $(6.18)_2$. However, the same values of $|\tilde{\Xi}|$ lead to different values of $|\tau_{\rm s}|$ if the size of microstructure is changed.

6.5 A Simple Gradient Theory

This section aims at a modification of the micromorphic grain boundary yield theory presented in Chap. 5 in order to numerically verify the theoretical investigations of the preceding sections. To this end, the grain boundary yield condition is modified according to Eq. (6.18). The grain boundary energy density W_Γ is chosen to vanish, i.e., the grain boundary yield strength is assumed to remain constant during the deformation process. The omission of an interface energy renders the microtraction $\Xi_\Gamma = [\![\boldsymbol{\xi}]\!] \cdot \boldsymbol{n}$ (Sect. 5.1.3, Box 1) purely dissipative , i.e., from Eq. (5.13) it follows that

$$\mathcal{D}_\Gamma = \Xi_\Gamma \, \dot{\zeta} \geq 0. \tag{6.19}$$

In Chap. 5, the following grain boundary yield condition has been applied (Eq. (5.19))

$$f_\Gamma = \Xi_\Gamma - \Xi_0^{\rm C}, \tag{6.20}$$

where $\Xi_0^{\rm C}$ represents the grain boundary yield strength. The necessary condition for grain boundary yielding reads $f_\Gamma = 0$. In this case, the

grain boundary loading Ξ_Γ reaches the critical value Ξ_0^{C}. Then, plastic grain boundary activity can occur, i.e. $\dot{\zeta} > 0$.
The criterion (6.20) may be applied to a periodic multi-grain microstructure with average grain diameter D. In this case, a Hall-Petch exponent of -1 is expected to describe the scaling behavior of the overall material response (see Eq. (6.6)).
However, in the following a different GB yield criterion than (6.20) will be applied. This is done to verify the theoretical comparison of the Eshelby and gradient models carried out in Sect. 6.4. If the theoretical investigations are correct, the Eshelby GB loading $|\tau_{\mathrm{s}}|$ may be roughly estimated by

$$|\tau_{\mathrm{s}}| = \frac{4k_1}{\sqrt{D}} \Xi_\Gamma, \tag{6.21}$$

compare Eq. (6.18). Here, k_1 has a slightly different interpretation. It might also depend on the grain misorientation. However, this dependence is neglected in the following.
As already mentioned, relation (6.21) can be verified by applying the yield condition of the Eshelby model to the discretized gradient plasticity model. This means that, instead of the yield condition (6.20), the condition

$$f_\Gamma^* = \frac{1}{4k_1}(|\tau_{\mathrm{s}}| - \tau_{\mathrm{s}}^{\mathrm{C}}) \leq 0 \tag{6.22}$$

will be used for the subsequent gradient plasticity simulation. The value of $|\tau_{\mathrm{s}}|$ is estimated based on Eq. (6.21). As a consequence, the scaling behavior of the gradient model is expected to be consistent with the Eshelby model.

Remark: A major motivation of the introduction of gradient theories is the desire to model size effects. Clearly, it is not desirable to formulate constitutive equations which depend on the characteristic system length scale. Instead, all parameters entering the constitutive equations should be constant material parameters. Eqns. (6.21) and (6.22) do not meet this requirement since the grain size explicitly enters Eq. (6.21). Therefore, it is emphasized that this model is subsequently applied exclusively to verify the theoretical comparison of the Eshelby and the gradient model, as outlined in Sect. 6.4.2 (in particular Eq. (6.18)).

6.6 Numerical Simulations

6.6.1 Simulation Set-Up

The numerical implementation of the theory is based on a micromorphic approximation which is documented in detail in Sect. 5.2. For the isotropic hardening, the Voce hardening relation (5.34) is applied. The elastic energy is taken as $W_{\mathrm{e}} = \boldsymbol{\varepsilon}^{\mathrm{e}} \cdot \mathbb{C}[\boldsymbol{\varepsilon}^{\mathrm{e}}]/2$ and the defect energy W_{g} is represented by the quadratic approach

$$W_{\mathrm{g}}(\nabla\zeta) = \frac{1}{2} K_{\mathrm{G}} \nabla\zeta \cdot \nabla\zeta. \tag{6.23}$$

Finally, the bulk flow rule (5.18) as well as the grain boundary yield condition, given by Eqns. (6.21) and (6.22), are applied. Table 6.1 summarizes the material parameters. An interpretation of the material parameters can be found in App. C.3.

Table 6.1: Model parameters

C_{1111}	C_{1122}	C_{1212}	K_{G}	p	$k^{\mathrm{C}} = \tau_{\mathrm{s}}^{\mathrm{C}}/(4k_1)$
168 GPa	121 GPa	75 GPa	0.01 N	200	13.05 MPa µm$^{-1/2}$
$\dot{\gamma}_0$	Θ_0	τ_0^{C}	τ_∞^{C}	τ^{D}	
10^{-3} 1/s	160 MPa	5 MPa	100 MPa	5 MPa	

A full periodic Voronoi microstructure consisting of 422 grains models the polycrystal. The RVE is loaded in tension in x_1-direction with periodic boundary conditions and free lateral contractions. This means that the volume averages of the lateral normal stresses $\bar{\sigma}_{22}$ and $\bar{\sigma}_{33}$ vanish. The FEM-mesh is based on linear hexahedrons and has ~1 million degrees of freedom. The first solution time steps are depicted in Fig. 6.1.

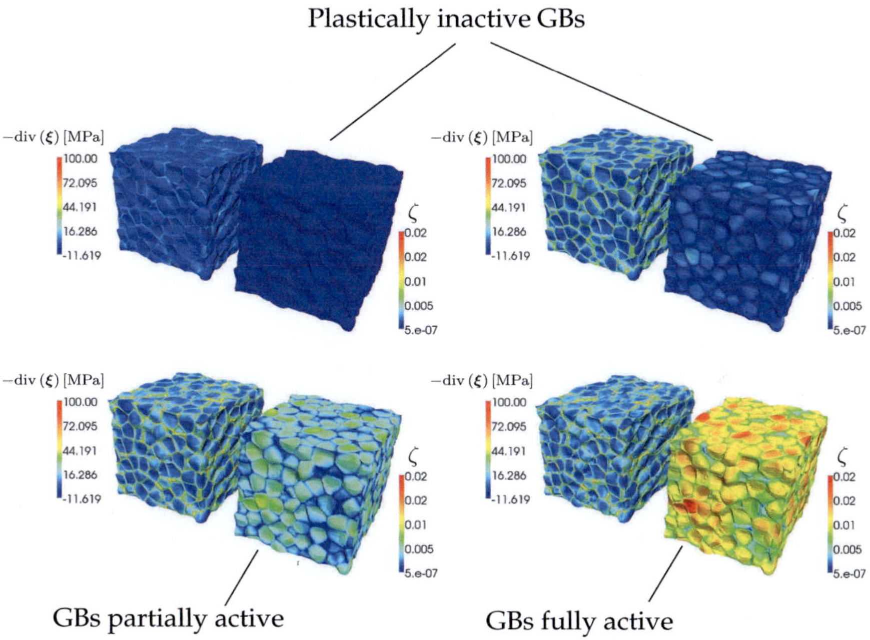

Figure 6.1: Tensile test simulation of a periodic Voronoi RVE with 422 grains. The figures illustrate the micromorphic counterpart $\zeta \approx \gamma_{\mathrm{eq}}$ of the equivalent plastic strain and the stress $-\mathrm{div}\,(\boldsymbol{\xi})$. In addition, the transition from plastically inactive to plastically active grain boundaries is shown.

6.6.2 Interpretation

The simulation results clearly illustrate that the onset of the plastic deformation is located inside of the grains, while the grain boundaries remain plastically inactive during the initial phase. The plastic grain boundary deformation is delayed since the grain boundary yield condition $f_\Gamma^* = 0$ is not instantly satisfied.

When the grain boundary jumps of $\nabla\zeta$ and the associated microstresses $\boldsymbol{\xi} = K_{\mathrm{G}} \nabla\zeta$ become sufficiently large, the grain boundary stress $|\tau_{\mathrm{s}}| = 4k_1\, \Xi_\Gamma/\sqrt{D} = 4k_1\, [\![\boldsymbol{\xi}]\!] \cdot \boldsymbol{n}/\sqrt{D}$ reaches the critical value $\tau_{\mathrm{s}}^{\mathrm{C}}$ and the grain boundary yields. This effect models dislocation pile-ups which build up at the grain boundaries (see Fig. 6.1).

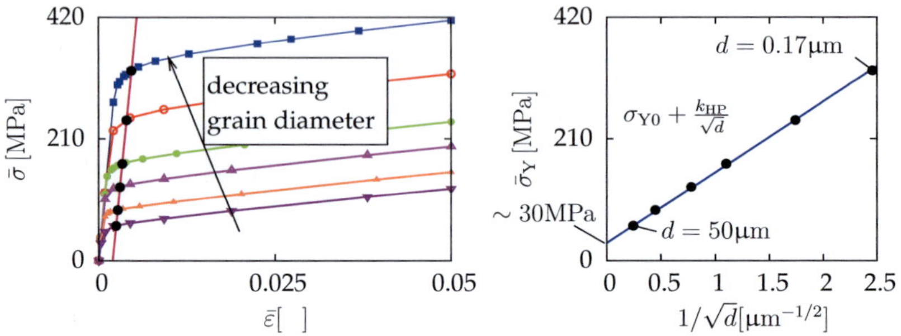

Figure 6.2: Simulated macroscopic stress-strain diagrams. The simulated 0.2% offset yield stresses are indicated by black filled circles. The right figure compares the simulation results to the Hall-Petch relation, Eq. (6.1).

Once, the grain boundary break through is reached (i.e. the grain boundary yields) the growth rate of the gradient stress $\mathrm{div}(\boldsymbol{\xi}) = K_G \Delta\zeta$ reduces significantly. A possible interpretation of this result is related to the dislocation pile-up density roughly remaining constant after the onset of grain boundary yielding.

The simulation has been carried out for different cube sizes but geometrically similar microstructures. The stress strain curves (Fig. 6.2, left) show the typical Hall-Petch behavior, i.e., the macroscopic yield stress σ_Y depends on the average grain size D, while the macroscopic hardening slope barely shows any size dependence.

Figure 6.2 (right) illustrates a comparison of the simulated 0.2% offset yield stress and the Hall-Petch relation (6.1)

$$\sigma_Y = \sigma_{Y0} + \frac{k_{HP}}{\sqrt{D}}. \tag{6.24}$$

For large grain sizes, the extrapolated yield stress comes out to be $\sigma_{Y0} \approx 30$ MPa which matches the theoretical prediction based on the Taylor factor (see App. C.3, Eq. (C.18))

$$M_T(\tau_0^C + \tau^D) = 30.6 \text{ MPa}. \tag{6.25}$$

The simulated Hall-Petch slope reads $k_{\mathrm{HP}} \approx 122\,\mathrm{MPa}\,\mu\mathrm{m}^{-1/2}$, which is also close to the prediction proposed in App. C.3, Eq. (C.22)

$$k_{\mathrm{HP}} \approx 3k_{\varepsilon}\frac{\tau_{\mathrm{s}}^{\mathrm{C}}}{4k_1} = 117.45\,\mathrm{MPa}\,\mu\mathrm{m}^{-1/2}. \tag{6.26}$$

The good match of the simulation results with the Hall-Petch line (Fig. 6.2) confirms the theoretical comparison of the Eshelby model and gradient plasticity models outlined in Sect. 6.4. It is emphasized again, that the good match is a result of the rescaling of the gradient plasticity grain boundary loading (i.e. the division of Ξ_{Γ} by $\sqrt{D}$, Eq. (6.21)). The gradient theory is unable to match the Hall-Petch exponent of 0.5 without this rescaling.

Chapter 7

Revision of a Continuum Dislocation Dynamics Theory at Small Strains

The principal motivation of this chapter is the revision of a dislocation continuum theory, which can extend the gradient plasticity framework developed in Chaps. 4 and 5. This extension allows for a significantly more detailed kinematical description of the dislocation microstructure in the single crystal model as well as the formulation of associated hardening models.

The theory is a higher-dimensional kinematical dislocation framework, developed by Hochrainer (2006), and can be represented by two partial differential equations. These two equations govern the evolution of the two field variables of the theory, given by the space-, time- and orientation-dependent dislocation density and curvature. The theory can be applied to systems of three-dimensional, continuously distributed, plane and curved dislocations. The evolution equations were derived by Hochrainer from the evolution equation of a higher-dimensional dislocation-density tensor – a generalization of Nye's tensor.

Aiming at contributing physical interpretations of the theory, the work at hand suggests how each of the terms of the evolution equations of Hochrainer (2006) can be motivated individually from the kinematics of individual dislocations. In addition, a simplified version of the higher-dimensional theory (Hochrainer et al., 2010) is discussed, which is numerically more attractive.

In Chap. 8, the simplified theory outlined in this chapter is coupled to the gradient plasticity theory of Chap. 5, yielding a dislocation based crystal plasticity theory. The resulting model contains a considerable amount of information on the dislocation microstructure. At the same time, it remains (at least in certain cases) computationally manageable.

7.1 Introduction

Hochrainer (2006) generalized Nye's dislocation density tensor to a kinematical framework for a higher-dimensional description of continuously distributed, curved dislocations. Hochrainer's dislocation kinematics can be regarded as the basis for a Continuum Dislocation Dynamics Theory (CDD). The introduction of the dislocation orientation as additional dimension allows the definition of a higher-dimensional dislocation density tensor which contains more detailed information on the dislocation microstructure than Nye's dislocation density tensor. Amongst other aspects, it allows for the computation of the geometrically necessary dislocation density and the total dislocation density.
Hochrainer and Zaiser (2005) showed that for plane dislocations, the tensor is completely described by two scalar higher-dimensional fields, namely the orientation-dependent dislocation density and the average curvature. Moreover, they generalized the evolution equation of Nye's tensor for a given dislocation velocity to the higher-dimensional counterpart and thus derived the evolution equations of the aforementioned dislocation density and curvature. The equations were numerically solved and constitutively coupled with the stress field of a bending specimen by Sandfeld et al. (2010) based on the finite difference method. Here, the focus lies on the evolution equation of the orientation dependent dislocation density derived by Hochrainer (2006). Aiming at contributing interpretations of the different terms of the equations, each of the terms is separately motivated based on the kinematics of discrete dislocations. Similarities between the transport equation of continuously distributed mass density (i.e. the continuity equation) and the spatial dislocation transport term of the dislocation evolution equation are discussed (e.g. Hochrainer et al., 2007).

Subsequently, the production of new dislocation line length as a result of the motion of a single dislocation is illustrated. This process is generalized in Hochrainer's concept of continuously distributed orientated and curved dislocations.
Finally, the work concentrates on the evolution of dislocation segments, the ends of which are constrained to have constant orientations. All dislocations except those segments with a particular orientation are blinded out. This consideration allows to describe how – in the case of an orientation-dependent dislocation density – the transport from one orientation to the adjacent orientation can be exploited to reproduce the associated term of Hochrainer's theory.

7.2 Kinematics of a Discrete Dislocation

A single slip system is considered. The slip direction and slip plane normal are given by $\boldsymbol{d} = \boldsymbol{e}_1$ and $\boldsymbol{n} = \boldsymbol{e}_3$, respectively (where $\{\boldsymbol{e}_1, \boldsymbol{e}_2, \boldsymbol{e}_3\}$ is a right-handed orthonormal system) and dislocation climb is neglected. Planar dislocations and small deformations are considered exclusively, such that several slip systems can be modeled by means of superposition.
An arbitrary discrete dislocation with index i is represented by a smooth curve (see Fig. 7.1)

$$\mathcal{C}_i: \quad \boldsymbol{x}_i = \boldsymbol{x}_i(S_i, t), \tag{7.1}$$

where S_i is the arc length in the initial state at $t = 0$ and $s_i = s_i(S_i, t)$ denotes the arc length in the current configuration.
Additionally, the curve can be described in the current configuration by $\tilde{\boldsymbol{x}}_i(s_i, t) = \boldsymbol{x}_i(S_i(s_i, t), t)$, given that the inverse $S_i(s_i, t) = s_i^{-1}(s_i, t)$ exists and is unique. In the following, the motion of two families of points will be of special relevance.

- Points $\boldsymbol{x}_i(S_i = S_{i,0}, t) =: \boldsymbol{y}_i(t)$ with a constant arc length S_i in the initial configuration will be called *quasi-material points*. They coincide with actual material points, if the motion of a dislocation is mimicked by a material object, for example a rubber band, the center-line of which is to follow the same motion $\boldsymbol{x}_i(S_i,t)$ as the dislocation.

- Points which are constrained to lie on the dislocation, may "slide" along it and thereby do not represent quasi-material objects. The direction of the tangents at these points will be restricted to be constant (see point P in Fig. 7.1). Consequently, they will travel along the dislocation if it rotates locally.

In the following, the dislocation motion is assumed to be such that the velocity of the quasi-material points

$$\boldsymbol{\nu}_i(S_i,t) = \partial_t \boldsymbol{x}_i(S_i,t) = \nu_i \boldsymbol{e}_{\nu,i} \quad \text{(no summation)} \tag{7.2}$$

is perpendicular to the line direction

$$\boldsymbol{e}_{\mathrm{l},i}(S_i,t) = \frac{1}{\|\partial_{S_i}\boldsymbol{x}_i(S_i,t)\|}\partial_{S_i}\boldsymbol{x}_i(S_i,t). \tag{7.3}$$

In that case the inverse $S_i(s_i,t)$ is unique, such that $\boldsymbol{e}_{\mathrm{l},i} = \partial_{s_i}\boldsymbol{x}_i(s_i,t)$.

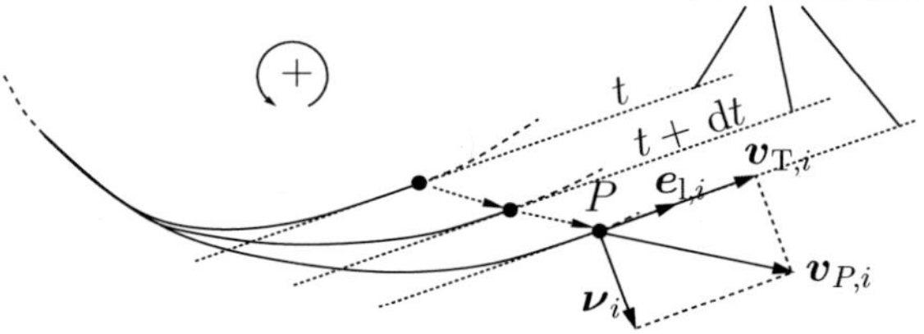

Figure 7.1: Representation of a dislocation by a smooth curve $\mathcal{C}_i$. The point P is sliding along the dislocation such that the tangent in P is constant. For illustration reasons the left end is frozen.

The angular velocity is defined by (e.g. Sandfeld et al., 2010)

$$\vartheta_i(s_i,t) = -\partial_{s_i}\nu_i(s_i,t). \tag{7.4}$$

The motion of an infinitesimal quasi-material segment, identified by its initial arc length S_i, is described by $\boldsymbol{\nu}_i(S_i,t)$ and $\vartheta_i(s_i(S_i,t),t)$, i.e., the state of the segment is a point in the Euclidean space extended by the orientation space. It can be shown (see App. D) that the velocity of a point P at s_i in $\boldsymbol{e}_{1,i}$-direction is given by

$$\boldsymbol{v}_{\mathrm{T},i}(s_i, t) = -\frac{\vartheta_i}{k_i}\boldsymbol{e}_{1,i}, \tag{7.5}$$

if the point is constrained to lie on the dislocation and the direction of the tangent at the point is restricted to be constant (see Fig. 7.1). Here $k_i = k_i(s_i, t)$ is the local (signed) curvature.
The total velocity $\boldsymbol{v}_{P,i}$ of the point P is then given by

$$\boldsymbol{v}_{P,i}(s_i, t) = \boldsymbol{\nu}_i(s_i, t) + \boldsymbol{v}_{\mathrm{T},i}(s_i, t) = \nu_i \boldsymbol{e}_{\nu,i} - \frac{\vartheta_i}{k_i}\boldsymbol{e}_{1,i}. \tag{7.6}$$

7.3 Review of the Continuity Equation

The derivation of the evolution equation of the dislocation density will be based on an analogon. First, the transfer of a model of discrete particles, for example atoms or molecules, to the continuity equation of a continuously distributed mass will be reviewed.
Let a given volume of interest Ω contain a large number of particles and let an arbitrary subvolume $\Delta\Omega \subset \Omega$ with $|\Delta\Omega| \ll |\Omega|$ and centroid $\boldsymbol{x}$. It is assumed that $\Delta\Omega$ contains a sufficiently large number of particles $\tilde{N} \gg 1$. If the dimensions of the particles are negligible they can be treated like point-masses (see Fig. 7.2).
The total particle mass (divided by $|\Delta\Omega|$) contained in $\Delta\Omega$ is given by

$$\frac{\Delta m}{|\Delta\Omega|} = \frac{1}{|\Delta\Omega|}\sum_{i=1}^{\tilde{N}} m_i, \tag{7.7}$$

where m_i is the mass of the i-th particle.
In continuum mechanics the counterpart of the total particle mass Δm divided by $\Delta\Omega$ is the mass density $\rho_m(\boldsymbol{x}, t)$, which represents

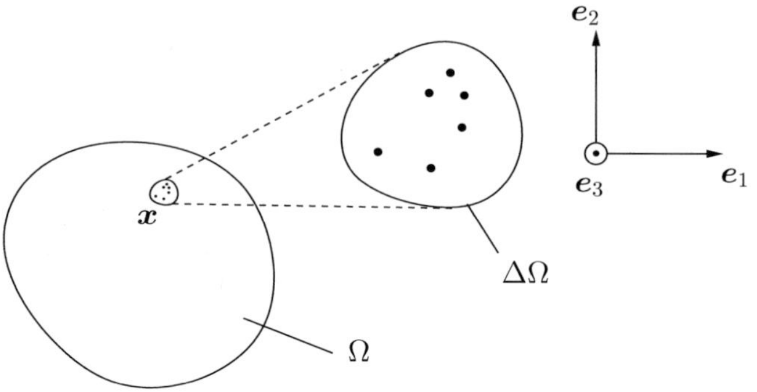

Figure 7.2: Subvolume $\Delta\Omega$ containing $\tilde{N} \gg 1$ particles

the "continuously distributed mass per unit volume". The subset of particles in $\Delta\Omega$ is associated with a mass flux (the linear momentum divided by $|\Delta\Omega|$)

$$\boldsymbol{f}_m = \frac{1}{\Delta\Omega} \sum_{i=1}^{\tilde{N}} m_i \boldsymbol{v}_i, \tag{7.8}$$

where $\boldsymbol{v}_i$ is the spatial velocity of particle i.
The corresponding flux of a continuously distributed mass is $\rho_m \boldsymbol{v}$, where $\boldsymbol{v} = \boldsymbol{v}(\boldsymbol{x}, t)$ is the velocity field. The mass flux through a given area A with normal $\tilde{\boldsymbol{n}}$ is $\int_A \rho_m \boldsymbol{v} \cdot \tilde{\boldsymbol{n}} \, \mathrm{d}a$.
The balance equation of the mass – the continuity equation – states that the time derivative of the total mass in an arbitrary fixed volume V is given by the negative outward mass flux

$$\partial_t \int_V \rho_m \, \mathrm{d}v = - \int_{\partial V} \rho_m \boldsymbol{v} \cdot \tilde{\boldsymbol{n}} \, \mathrm{d}a. \tag{7.9}$$

Note that, due to the conservation of mass, this equation comprises neither a production nor a supply term.

7.4 An Orientation-Dependent Dislocation Density

In the following, the modeling process presented for point masses is adopted for dislocation lines, aiming at an interpretation of the "Continuum Dislocation Dynamics"-theory of Hochrainer et al. (2007), who derived an evolution equation of an orientation-dependent dislocation density. Interpretations of all terms of the evolution equation are given by motivating them – one after another – from their respective discrete counterpart (a finite number of dislocations).

Once more, let the volume Ω (a single crystal) contain a huge number of dislocations and let the subvolume $\Delta\Omega \subset \Omega$ with $|\Delta\Omega| \ll |\Omega|$ contain a sufficiently large number of dislocations. Here, the special case where all dislocations stay in their glide planes is considered. All dislocations are assumed to either represent closed curves or to end at the boundary $\partial\Omega$. At a given time t and a fixed but arbitrary orientation $\varphi \in [0, 2\pi)$, a set of points on the dislocation lines is considered, which is characterized by a common tangent direction $\boldsymbol{e}_l(\varphi) = \cos(\varphi)\boldsymbol{e}_1 + \sin(\varphi)\boldsymbol{e}_2$ (see Fig. 7.3).

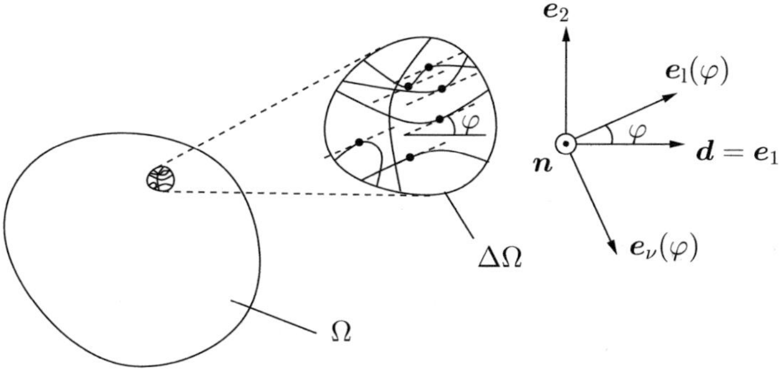

Figure 7.3: Discrete dislocations and tangent points with line orientation $\boldsymbol{e}_l(\varphi)$. Note that points of opposite line orientation $\boldsymbol{e}_l(\varphi + \pi)$ are not considered.

Let N_φ be the number of tangent points in $\Delta\Omega$ [7]. The index φ indicates that N_φ depends on the arbitrarily chosen but fixed orientation φ.

The total radius of curvature (divided by $|\Delta\Omega|$) is introduced by

$$\frac{\Delta R(\varphi)}{|\Delta\Omega|} = \frac{1}{|\Delta\Omega|} \sum_{i=1}^{N_\varphi} |R_{\varphi,i}|, \tag{7.10}$$

where $R_{\varphi,i}$ is the radius of curvature at tangent point i.

Note the formal equivalence of (7.10) and the definition (7.7) of the

[7] The case of straight lines, where $N_\varphi \to \infty$ will be treated in the subsequent text.

total mass (divided by $|\Delta\Omega|$), which will be explored in the subsequent derivation. This equivalence motivates the introduction of the orientation-dependent field $\rho_\varphi(\boldsymbol{x}, \varphi, t)$ which represents the "continuously distributed total radius of curvature per unit volume" (akin to the mass density which represents the "continuously distributed mass per unit volume").

To derive the link between the total dislocation density ("the total dislocation line length per unit volume") and the density $\rho_\varphi(\boldsymbol{x}, \varphi, t)$, M_φ discrete orientations $\varphi_j = \frac{j}{M_\varphi} 2\pi$ are introduced, where $j = 1, \ldots, M_\varphi$. Equivalently, $\varphi_{j+1} = \varphi_j + \Delta\varphi$ ($j = 1, \ldots, M_\varphi - 1$) holds with $\Delta\varphi = 2\pi/M_\varphi$. Then, in the case of discrete dislocations, for each orientation φ_j, a set of N_{φ_j} tangent quasi-material points can be identified in $\Delta\Omega$ with radii $R_{\varphi_j,i}$ and total radius of curvature $\Delta R(\varphi_j)$. Moreover, in the case of continuously distributed dislocations, an associated orientation-dependent field $\rho_\varphi(\boldsymbol{x}, \varphi_j, t)$ exists. To each of these tangent quasi-material points a segment of a circle can be attached. The radius of the circle is chosen to be equal to the local radius of curvature and the opening angle is set to $\Delta\varphi$. Additionally, the mid-point of the segment is chosen to coincide with the quasi-material point, such that the direction of the circle segment coincides with the direction of the dislocation at the tangent point (see Fig. 7.4).

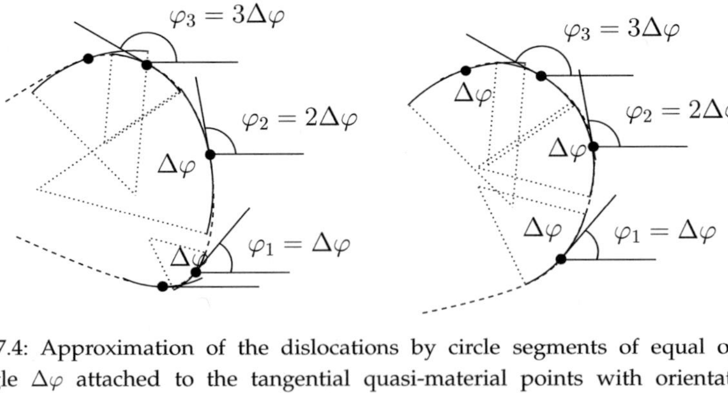

Figure 7.4: Approximation of the dislocations by circle segments of equal opening angle $\Delta\varphi$ attached to the tangential quasi-material points with orientations $\varphi_1, \varphi_2, \ldots, \varphi_{M_\varphi}$.

The union of all these discrete, discontinuous segments is an approximation of the smooth dislocation lines $\mathcal{C}_i$. For an infinitesimal opening angle $\Delta\varphi \to \mathrm{d}\varphi$, i.e., for $M_\varphi \to \infty$, the approximation converges to the curves $\mathcal{C}_i$.
The total line length of all segments associated with the orientation φ_j is given by

$$\Delta l_{\varphi_j} = \Delta\varphi \sum_{i=1}^{N_{\varphi j}} |R_{\varphi_j,i}| = \Delta\varphi \frac{1}{|\Delta\Omega|} \sum_{i=1}^{N_{\varphi j}} |R_{\varphi_j,i}||\Delta\Omega|, \tag{7.11}$$

where N_{φ_j} is the total number of tangent quasi-material points with orientation φ_j in $\Delta\Omega$. In the case of infinitesimal segments, Eq. (7.11) becomes

$$\mathrm{d}l_\varphi = \mathrm{d}\varphi \frac{1}{|\Delta\Omega|} \sum_{i=1}^{N_\varphi} |R_{\varphi,i}||\Delta\Omega|. \tag{7.12}$$

The associated expression in the case of continuously distributed dislocations is

$$\mathrm{d}l_\varphi = \rho_\varphi(\boldsymbol{x}, \varphi, t)\, \mathrm{d}\varphi\, \mathrm{d}v. \tag{7.13}$$

This result shows that the field $\rho_\varphi(\boldsymbol{x}, \varphi, t)$ can be interpreted as an orientation-dependent dislocation line length per unit volume *and* per unit angle (see also, e.g., Hochrainer, 2006). Consequently, it is referred to as *dislocation density*. The product $\rho_\varphi(\boldsymbol{x}, \varphi, t)\, \mathrm{d}\varphi$ represents the infinitesimal total line length per unit volume of continuously distributed dislocation segments with common orientation φ and opening angle $\mathrm{d}\varphi$. The total line length of all discrete segments in $\Delta\Omega$ is the sum of the contributions of all M_φ discrete orientations

$$\Delta l = \sum_{j=1}^{M_\varphi} \Delta l_{\varphi_j} \overset{(7.11)}{=} \left(\sum_{j=1}^{M_\varphi} \frac{1}{|\Delta\Omega|} \sum_{i=1}^{N_{\varphi j}} |R_{\varphi_j,i}|\, \Delta\varphi \right) |\Delta\Omega|. \tag{7.14}$$

The expression (7.14) motivates the definition of the total dislocation density $\rho(\boldsymbol{x}, t)$ in the continuous case (e.g. Hochrainer, 2006)

$$\mathrm{d}l = \rho(\boldsymbol{x}, t)\, \mathrm{d}v = \left(\oint \rho_\varphi(\boldsymbol{x}, \varphi, t)\, \mathrm{d}\varphi \right) \mathrm{d}v. \tag{7.15}$$

Note that ρ does not depend on the orientation. It represents the "total dislocation line length per unit volume". This result underlines the interpretation of the field $\rho_\varphi(\boldsymbol{x}, \varphi, t)$ as orientation-dependent "dislocation density per unit volume and per unit angle".
The case of straight dislocations can be handled using delta-distributions.

7.5 Dislocation Transport

The discrete flux of infinitesimal dislocation segments with orientation φ (divided by $|\Delta\Omega|$) is defined in analogy to the mass flux (Eq. (7.8))

$$\boldsymbol{f}(\varphi)\,\mathrm{d}\varphi = \left(\frac{1}{|\Delta\Omega|} \sum_{i=1}^{N_\varphi} |R_{\varphi,i}| \boldsymbol{\nu}_{\varphi,i} \right) \mathrm{d}\varphi. \tag{7.16}$$

In the following, a simplifying assumption is adopted, stating that the motion of the discrete dislocations is described by a smooth velocity field $\boldsymbol{\nu} = \boldsymbol{\nu}(\boldsymbol{x}, \varphi, t) = \nu(\boldsymbol{x}, \varphi, t)\boldsymbol{e}_\nu(\varphi)$ (where $\nu = |\boldsymbol{\nu}|$) in the sense that $\boldsymbol{\nu}_i(s_i, t) = \boldsymbol{\nu}(\boldsymbol{x}_i(s_i, t), \varphi_i(s_i, t), t)$, where $\varphi_i(s_i, t)$ (a smooth function of (s_i, t)) is the angle between the line direction $\boldsymbol{e}_{l,i}(s_i, t) = \boldsymbol{e}_l(\varphi_i(s_i, t))$ and the slip direction $\boldsymbol{d} = \boldsymbol{e}_1$.
Accordingly, all tangent quasi-material points with common orientation φ in the small volume $\Delta\Omega$ centered at $\boldsymbol{x}$ share approximately the same velocity $\boldsymbol{\nu}(\boldsymbol{x}, \varphi, t)$ (as $\Delta\Omega$ is considered to be sufficiently small, the Taylor series expansion of $\boldsymbol{\nu}$ around $\boldsymbol{x}$ can be truncated after the first, constant term). Consequently, the flux is given by (see Eq. (7.16))

$$\boldsymbol{f}(\varphi)\,\mathrm{d}\varphi \approx \frac{1}{|\Delta\Omega|} \sum_{i=1}^{N_\varphi} |R_{\varphi,i}| \boldsymbol{\nu}(\boldsymbol{x}, \varphi, t)\,\mathrm{d}\varphi. \tag{7.17}$$

This result motivates the definition of the orientation-dependent flux

$$\rho_\varphi(\boldsymbol{x}, \varphi, t)\boldsymbol{\nu}(\boldsymbol{x}, \varphi, t)\,\mathrm{d}\varphi \tag{7.18}$$

of the dislocation density $\rho_\varphi(\boldsymbol{x}, \varphi, t)$ in direction $\boldsymbol{e}_\nu(\varphi)$ in analogy to the mass density flux $\rho_m \boldsymbol{v}$. Eq. (7.18) is the continuous counterpart of (7.17).

The (infinitesimal) flux of continuously distributed dislocation segments with infinitesimally small opening angle $\mathrm{d}\varphi$ and orientation φ through a given area A can then be postulated to be

$$\left(\int\limits_A \rho_\varphi(\boldsymbol{x},\varphi,t)\boldsymbol{\nu}(\boldsymbol{x},\varphi,t)\cdot\tilde{\boldsymbol{n}}\,\mathrm{da}\right)\mathrm{d}\varphi. \tag{7.19}$$

It represents the transport of infinitesimal dislocation line length through that area in direction $\boldsymbol{e}_\nu(\varphi)$.

Note that these results depend crucially on the arbitrarily chosen and fixed orientation φ and on the assertion that all tangent points share the same velocity direction $\boldsymbol{e}_\nu(\varphi)$.

Moreover, the introduction of an infinitesimal time increment $\mathrm{d}t$ resulting in the time point $t+\mathrm{d}t$ may show that the tangents at the considered quasi-material points are no longer parallel, as their angular velocities are, in general, different.

Hence, at $t+\mathrm{d}t$ the tangent points associated to the fixed but arbitrary orientation φ will be given by a different set of quasi-material points adjacent to the quasi-material points considered at t with different radii of curvature $R_{\varphi,i}$.

Accordingly, the rate of change of the infinitesimal small total line length of the continuously distributed dislocation segments with common orientation φ and opening angle $\mathrm{d}\varphi$ in an arbitrary fixed volume V is given by

$$\begin{aligned}&\partial_t\left(\int\limits_V \rho_\varphi(\boldsymbol{x},\varphi,t)\,\mathrm{d}v\right)\mathrm{d}\varphi \qquad (7.20)\\ =&-\left(\int\limits_{\partial V}\rho_\varphi(\boldsymbol{x},\varphi,t)\nu(\boldsymbol{x},\varphi,t)\boldsymbol{e}_\nu(\varphi)\cdot\tilde{\boldsymbol{n}}\,\mathrm{da}\right)\mathrm{d}\varphi+\left(\int\limits_V w(\boldsymbol{x},\varphi,t)\,\mathrm{d}v\right)\mathrm{d}\varphi\end{aligned}$$

The production term $w=w(\boldsymbol{x},\varphi,t)$ accounts for the fact that at $t+\mathrm{d}t$ the local radii of curvature have changed with respect to t and is subject of the subsequent developments.

By virtue of the arbitrariness of V, the local form reads

$$\partial_t\rho_\varphi=-\mathrm{div}\,(\rho_\varphi\boldsymbol{\nu})+w, \tag{7.21}$$

where the arguments have been omitted.

7.6 Line Length Production

For the derivation of the source term $w(\boldsymbol{x}, \varphi, t)$ the curvature of all segments in $\Delta\Omega$ of an arbitrary but fixed orientation φ is assumed positive, for simplicity. The more general case, where different signs are allowed, will be considered later.

First, the start and end points of an arbitrarily chosen finite segment $\mathcal{L}(t)$ of the dislocation $\mathcal{C}_i$ are parameterized by the associated constant arc lengths S_i^{s} and S_i^{e} in the initial configuration. Then $s_i^{\mathrm{s}} = s_i(S_i^{\mathrm{s}}, t)$ and $s_i^{\mathrm{e}} = s_i(S_i^{\mathrm{e}}, t)$ are their analogon in the current configuration, giving rise to the definition $\mathcal{L}(t) = \{\boldsymbol{x}_i(s_i, t) : \ s_i \in [s_i^{\mathrm{s}}; s_i^{\mathrm{e}}],\ s_i^{\mathrm{s}} = s_i(S_i^{\mathrm{s}}, t);\ s_i^{\mathrm{e}} = s_i(S_i^{\mathrm{e}}, t)\}$.

The partial derivative of the arc length in the current configuration with respect to the initial configuration is given by

$$F_{S,i}(S_i, t) = \partial_{S_i} s_i. \tag{7.22}$$

Note the formal equivalence to the deformation gradient in continuum mechanics of deformable solids. The quantity $E_{S,i} = F_{S,i} - 1$ is interpreted as a quasi-material Biot-type strain measure which represents the relative elongation of a quasi-material segment $\mathrm{d}S_i$ with quasi-material end points

$$E_{S,i}(S_i, t) = \frac{\mathrm{d}s_i - \mathrm{d}S_i}{\mathrm{d}S_i}. \tag{7.23}$$

As the total length of the considered segment is given by $|\mathcal{L}| = l_{\mathcal{L}} = \int_{s_i^{\mathrm{s}}}^{s_i^{\mathrm{e}}} \mathrm{d}s_i$, its quasi-material time rate of change is defined by

$$\mathring{l}_{\mathcal{L}} = \frac{\partial}{\partial t} \int_{S_i^{\mathrm{s}}}^{S_i^{\mathrm{e}}} F_{S,i}\, \mathrm{d}S_i = \int_{s_i^{\mathrm{s}}}^{s_i^{\mathrm{e}}} (\partial_t F_{S,i}) F_{S,i}^{-1}\, \mathrm{d}s_i = \int_{s_i^{\mathrm{s}}}^{s_i^{\mathrm{e}}} d_{s,i}\, \mathrm{d}s_i \tag{7.24}$$

with the inverse $F_{S,i}^{-1} = \partial_{s_i} S_i$ of $F_{S,i}$ and the stretch rate $d_{s,i}(s_i, t) = (\partial_t F_{S,i})\, F_{S,i}^{-1}$.

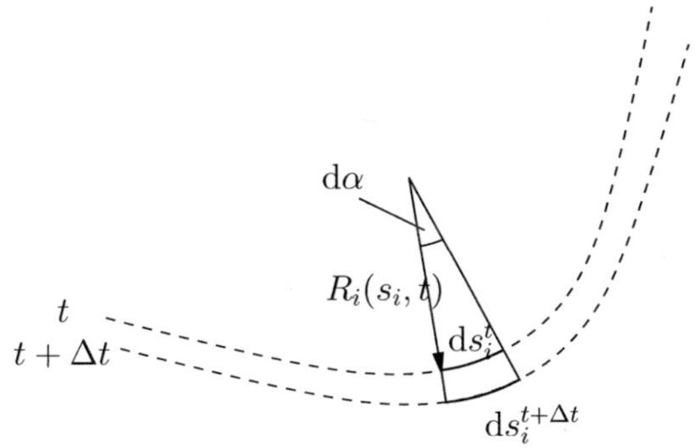

Figure 7.5: Incremental motion of a quasi-material segment with infinitesimal opening angle $\mathrm{d}\alpha$ at time t and $t + \mathrm{d}t$.

From Fig. 7.5 the arc length increment of a quasi-material infinitesimal segment at two subsequent time steps t and $t + \Delta t$ with small time increment Δt can be approximated

$$\mathrm{d}s_i^{t+\Delta t} - \mathrm{d}s_i^t \approx (R_i(s_i,t) + \nu_i(s_i,t)\Delta t)\,\mathrm{d}\alpha - R_i(s_i,t)\,\mathrm{d}\alpha = \nu_i(s_i,t)\Delta t\,\mathrm{d}\alpha. \tag{7.25}$$

In the limit case $\Delta t \to \mathrm{d}t$ the relation (7.25) becomes exact and it follows that

$$d_{s,i}\,\mathrm{d}t = \frac{\mathrm{d}s_i^{t+\,\mathrm{d}t} - \mathrm{d}s_i^t}{\mathrm{d}s_i} \Rightarrow d_{s,i}(s_i,t) = \frac{\nu_i(s_i,t)}{R_i(s_i,t)}. \tag{7.26}$$

7.7 Rotational Density Transport

In the next step, a more complex motion of a finite non-quasi-material dislocation segment $\mathcal{L}_{\Delta\varphi}(t) \subset \mathcal{C}_i$ is considered, like in Fig. 7.1. The end points of the segment are constrained to have the arbitrary but fixed orientation $\tilde{\varphi}$ (left end in Fig. 7.1) and $\tilde{\varphi} + \Delta\varphi$ (right end), respectively. They define $\mathcal{L}_{\Delta\varphi}(t) = \{\boldsymbol{x}_i(s_i,t) : s_i \in [s_i^{\mathrm{s}}; s_i^{\mathrm{e}}],\ \boldsymbol{e}_{\mathrm{l},i}(s_i^{\mathrm{s}},t) = \boldsymbol{e}_{\mathrm{l}}(\tilde{\varphi});\ \boldsymbol{e}_{\mathrm{l},i}(s_i^{\mathrm{e}},t) = \boldsymbol{e}_{\mathrm{l}}(\tilde{\varphi} + \Delta\varphi)\}$. Hence, the ends of the segment can, in general, not be identified by quasi-material points.

The angle $\Delta\varphi$ is required to be sufficiently small, such that the finite

segment can be approximated by a segment of a circle and the inverse $s_i(\varphi_i,t)$ of $\varphi_i(s_i,t)$ can be defined on $\mathcal{L}_{\Delta\varphi}$. Locally, the radius of curvature and angular velocity can be expressed as $\tilde{R}_i(\varphi_i,t) = R_i(s_i(\varphi_i,t),t)$ and $\tilde{\vartheta}_i(\varphi_i,t) = \vartheta_i(s_i(\varphi_i,t),t)$, $s_i \in [s_i^{\mathrm{s}}; s_i^{\mathrm{e}}]$, respectively.
According to Fig. 7.1, the line length of segment $\mathcal{L}_{\Delta\varphi}(t)$ increases or decreases due to a nonzero angular velocity $\vartheta_i(s_i^{\mathrm{s}},t)$ and $\vartheta_i(s_i^{\mathrm{e}},t)$ at the ends. According to Fig. 7.1 and Eq. (7.5), the associated line-"production" at $\varphi + \Delta\varphi$ during a time interval $\mathrm{d}t$ is

$$v_{\mathrm{T},i}(s_i^{\mathrm{e}},t)\,\mathrm{d}t \;=\; -R_i(s_i^{\mathrm{e}},t)\vartheta_i(s_i^{\mathrm{e}},t)\,\mathrm{d}t = -\tilde{R}_i(\varphi+\Delta\varphi,t)\tilde{\vartheta}_i(\varphi+\Delta\varphi,t)\,\mathrm{d}t.$$

The total rate of change of the segment length $|\mathcal{L}(t)| = l_{\Delta\varphi} = s_i^{\mathrm{e}} - s_i^{\mathrm{s}}$ is given by the sum of the quasi-material time rate of change of $l_{\Delta\varphi}$ and the "production"-rates at the ends

$$\frac{\mathrm{d}l_{\Delta\varphi}}{\mathrm{d}t} = \mathring{l}_{\Delta\varphi} + v_{\mathrm{T},i}(s_i^{\mathrm{e}},t) - v_{\mathrm{T},i}(s_i^{\mathrm{s}},t) \overset{(7.24),(7.26)}{=} \int\limits_{s_i^{\mathrm{s}}}^{s_i^{\mathrm{e}}} \frac{\nu_i}{R_i}\,\mathrm{d}s_i - \tilde{R}_i\tilde{\vartheta}_i\Big|_{\varphi+\Delta\varphi} + \tilde{R}_i\tilde{\vartheta}_i\Big|_{\varphi}. \tag{7.27}$$

For any infinitesimal subsegment of $\mathcal{L}_{\Delta\varphi}(t)$ at $s_i \in [s_i^{\mathrm{s}}; s_i^{\mathrm{e}}]$ (or $\varphi \in [\tilde{\varphi}; \tilde{\varphi}+\Delta\varphi]$, respectively) with length $\mathrm{d}s_i = R_i(s_i,t)\,\mathrm{d}\varphi$, the infinitesimal limiting case of Eq. (7.27) reads

$$\mathrm{d}\varphi\frac{\mathrm{d}}{\mathrm{d}t}R_i(s_i,t) = \mathrm{d}\varphi\nu_i(s_i,t) - \partial_\varphi\big(\tilde{R}_i(\varphi,t)\tilde{\vartheta}_i(\varphi,t)\big)\,\mathrm{d}\varphi. \tag{7.28}$$

To carry this result over to continuously distributed dislocations, the angular velocity is considered

$$\begin{aligned}
\vartheta_i(s_i,t) &= \partial_{s_i}\nu_i(s_i,t) \\
&= \nabla\nu(\boldsymbol{x}_i(s_i,t),\varphi_i(s_i,t),t)\cdot\partial_{s_i}\boldsymbol{x}_i + \big(\partial_\varphi(\nu(\boldsymbol{x}_i(s_i,t),\varphi_i(s_i,t),t))\big(\partial_{s_i}\varphi_i\big) \\
&= \nabla\nu(\boldsymbol{x}_i,\varphi_i,t)\cdot\mathbf{e}_{\mathrm{l}}(\varphi_i) + k_i(s_i,t)\,\partial_\varphi\nu(\boldsymbol{x}_i,\varphi_i,t) && (7.29)\\
&\approx \nabla\nu(\boldsymbol{x},\varphi_i,t)\cdot\mathbf{e}_{\mathrm{l}}(\varphi_i) + k_i(s_i,t)\,\partial_\varphi\nu(\boldsymbol{x},\varphi_i,t), && (7.30)
\end{aligned}$$

where k_i is the local radius of curvature.
This result leads to an interesting expression of the line average $\bar{\vartheta}_\varphi$ of the

angular velocity of all infinitesimal segments with common orientation $\varphi_i = \varphi$ in $\Delta\Omega$

$$\begin{aligned} \bar{\vartheta}_\varphi &= \frac{\sum_i \vartheta_{\varphi,i}\,\mathrm{d}l_{\varphi,i}}{\sum_i \mathrm{d}l_{\varphi,i}} = \frac{\sum_i \vartheta_{\varphi,i} R_{\varphi,i}}{\sum_i R_{\varphi,i}} && (7.31)\\ &\overset{(7.30)}{=} \boldsymbol{e}_{\mathrm{l}}(\varphi)\cdot\nabla\nu(\boldsymbol{x},\varphi,t) + \frac{\sum_i k_{\varphi,i}\,\mathrm{d}l_{\varphi,i}}{\sum_i \mathrm{d}l_{\varphi,i}}\partial_\varphi \nu(\boldsymbol{x},\varphi,t) && (7.32)\\ &= \boldsymbol{e}_{\mathrm{l}}(\varphi)\cdot\nabla\nu(\boldsymbol{x},\varphi,t) + \bar{k}_\varphi\,\partial_\varphi\nu(\boldsymbol{x},\varphi,t), && (7.33)\end{aligned}$$

with the (orientation dependent) line average of the curvature

$$\bar{k}_\varphi := \frac{\sum_i k_{\varphi,i}\,\mathrm{d}l_{\varphi,i}}{\sum_i \mathrm{d}l_{\varphi,i}} = \frac{\sum_i k_{\varphi,i} R_{\varphi,i}}{\sum_i R_{\varphi,i}} \overset{(k_{\varphi,i}=1/R_{\varphi,i})}{=} \frac{N_\varphi}{|\Delta\Omega|}\frac{|\Delta\Omega|}{\sum_i R_{\varphi,i}} = \bar{n}_\varphi \frac{|\Delta\Omega|}{\sum_i R_{\varphi,i}}. \tag{7.34}$$

In (7.31) and (7.34), the sums run over all N_φ segments in $\Delta\Omega$. The number N_φ of infinitesimal dislocation segments with common orientation φ divided by $|\Delta\Omega|$

$$\bar{n}_\varphi = \frac{N_\varphi}{|\Delta\Omega|} \tag{7.35}$$

will play an important role in the subsequent developments. Therefore its continuously distributed counterpart is introduced. It will be called *distributed segment density* $n(\boldsymbol{x},\varphi,t)$ ("number of infinitesimal segments with orientation φ per unit volume"). The (line-)average curvature $k(\boldsymbol{x},\varphi,t)$ in the continuous case can then be defined by (see Eq. 7.34)

$$k(\boldsymbol{x},\varphi,t) = \frac{n(\boldsymbol{x},\varphi,t)}{\rho_\varphi(\boldsymbol{x},\varphi,t)}. \tag{7.36}$$

Motivated by Eq. (7.31) the line average of the angular velocity is defined in the continuous case by

$$\vartheta(\boldsymbol{x},\varphi,t) = \boldsymbol{e}_{\mathrm{l}}(\varphi)\cdot\nabla\nu(\boldsymbol{x},\varphi,t) + k(\boldsymbol{x},\varphi,t)\partial_\varphi\nu. \tag{7.37}$$

Relations equivalent to (7.36) and (7.37) can be found, e.g., in Hochrainer et al. (2007).

From Eqns. (7.28) and $\nu_i(s_i,t) \approx \nu(\boldsymbol{x},\varphi,t)$ (as $\Delta\Omega$ is small), the rate of change of the total line length of all N_φ segments in $\Delta\Omega$ follows by superposition

$$\frac{\mathrm{d}}{\mathrm{d}t}\left(\frac{1}{|\Delta\Omega|}\sum_{i=1}^{N_\varphi} R_{\varphi,i}\right)\mathrm{d}\varphi|\Delta\Omega| \tag{7.38}$$

$$= \frac{N_\varphi}{|\Delta\Omega|}\nu(\boldsymbol{x},\varphi,t)\,\mathrm{d}\varphi|\Delta\Omega| - \partial_\varphi\left[\left(\frac{1}{|\Delta\Omega|}\sum_{i=1}^{N_\varphi} R_{\varphi,i}\right)\underbrace{\left(\frac{\sum_i \vartheta_{\varphi,i}R_{\varphi,i}}{\sum_i R_{\varphi,i}}\right)}_{\bar{\vartheta}_\varphi}\right]\mathrm{d}\varphi|\Delta\Omega|,$$

where the segment flux over the boundary $\partial\Delta\Omega$ has been neglected.
An analogue expression in the continuous case is given by

$$w(\boldsymbol{x},\varphi,t)\,\mathrm{d}\varphi\,\mathrm{d}v = \big(n(\boldsymbol{x},\varphi,t)\nu(\boldsymbol{x},\varphi,t) - \partial_\varphi(\rho_\varphi(\boldsymbol{x},\varphi,t)\vartheta(\boldsymbol{x},\varphi,t))\big)\,\mathrm{d}\varphi\,\mathrm{d}v, \tag{7.39}$$

where $n, \nu, \rho_\varphi, \vartheta$ and, consequently, w are functions of $(\boldsymbol{x},\varphi,t)$. This result yields the evolution equation of the dislocation density ρ_φ (see Eq. (7.21))

$$\frac{\partial}{\partial t}\int_V\int_{\Delta\varphi}\rho_\varphi\,\mathrm{d}\varphi\,\mathrm{d}v = -\int_{\Delta\varphi}\int_{\partial V}\rho_\varphi\boldsymbol{\nu}\cdot\tilde{\boldsymbol{n}}\,\mathrm{d}a\,\mathrm{d}\varphi + \int_V\left[\int_{\Delta\varphi} n\nu\,\mathrm{d}\varphi - \left(\rho_\varphi\vartheta\Big|_{\varphi+\Delta\varphi} - \rho_\varphi\vartheta\Big|_{\varphi}\right)\right]\mathrm{d}v, \tag{7.40}$$

where the domain of integration $V\times[\varphi;\varphi+\Delta\varphi]$ is arbitrary.
The local form represents the evolution equation of the dislocation density

$$\partial_t\rho_\varphi + \operatorname{div}(\rho_\varphi\boldsymbol{\nu}) = -\partial_\varphi(\rho_\varphi\vartheta) + n\nu, \tag{7.41}$$

as can be found in, e.g., Hochrainer (2006). To solve this equation, it has to be supplemented with initial and boundary conditions. The evolution equation of n is subject of the subsequent developments.

7.8 Evolution of the Segment Density

The solution of Eq. (7.41) requires the evolution of the segment density $n(\boldsymbol{x},\varphi,t)$ which is the continuously distributed number of tangent points with common orientation φ per unit volume. The kinematics

of the associated tangent points, which are constrained to have constant orientation, has already been analysed. Their motion is exactly described by the trajectory of point P in Fig. 7.1.
The derivation of the evolution equation of $n(\boldsymbol{x}, \varphi, t)$ can again be postulated by analogy to the continuity equation of the mass density $\rho_m(\boldsymbol{x}, t)$. Therefore, note the formal equivalence of

$$\bar{n}_\varphi = \frac{N_\varphi}{\Delta\Omega} = \frac{1}{\Delta\Omega}\sum_{i=1}^{N_\varphi} 1 \tag{7.42}$$

and Eq. (7.7). Considering all N_φ segments in the small volume $\Delta\Omega$, the associated segment flux can be defined analogously (see Eqns. (7.6) and (7.8))

$$\begin{aligned}
\frac{1}{|\Delta\Omega|}\sum_i 1\,\boldsymbol{v}_{P,i} &= \frac{1}{|\Delta\Omega|}\sum_i (\boldsymbol{\nu}_{\varphi,i} + \boldsymbol{v}_{\mathrm{T}_{\varphi,i}}) \\
&\approx \frac{N_\varphi}{|\Delta\Omega|}\boldsymbol{\nu}(\boldsymbol{x}, \varphi, t) - \frac{1}{|\Delta\Omega|}\sum_i R_{\varphi,i}\vartheta_{\varphi,i}\boldsymbol{e}_1(\varphi) \\
&\overset{(7.31)}{=} \bar{n}_\varphi \boldsymbol{\nu}(\boldsymbol{x}, \varphi, t) - \frac{\sum_i R_{\varphi,i}}{|\Delta\Omega|}\bar{\vartheta}_\varphi \boldsymbol{e}_1(\varphi).
\end{aligned}$$

This result allows to define the flux in the case of continuously distributed dislocations associated to the segment density $n(\boldsymbol{x}, \varphi, t)$ by

$$n\boldsymbol{v}_P = n(\boldsymbol{x}, \varphi, t)\boldsymbol{\nu}(\boldsymbol{x}, \varphi, t) - \rho_\varphi(\boldsymbol{x}, \varphi, t)\vartheta(\boldsymbol{x}, \varphi, t)\boldsymbol{e}_1(\varphi) = n\left(\boldsymbol{\nu} - \frac{\vartheta}{k}\boldsymbol{e}_1\right), \tag{7.43}$$

which represents an implicit definition of the average segment velocity

$$\boldsymbol{v}_P = \boldsymbol{\nu} + \boldsymbol{v}_{\mathrm{T}} = \nu\boldsymbol{e}_\nu - \frac{\vartheta}{k}\boldsymbol{e}_1, \tag{7.44}$$

such that the segment flux is given by $n\boldsymbol{v}_P$. Note the remarkable formal equivalence to Eq. (7.6). As neither sources nor dislocation annihilation are considered, the production of n is zero and it follows that (Hochrainer, 2006; Hochrainer et al., 2007; Hochrainer, 2013)

$$\partial_t \int_V n \,\mathrm{d}v = -\int_{\partial V} n\boldsymbol{v}_P \cdot d\boldsymbol{a} \;\Rightarrow\; \partial_t n + \operatorname{div}(n\boldsymbol{v}_p) = \partial_t n + \operatorname{div}(n\boldsymbol{\nu} - \rho_\varphi\vartheta\boldsymbol{e}_1) = 0. \tag{7.45}$$

Eqns. (7.41) and (7.45) have been derived based on the assumption of strictly positive curvatures. In the more general case, where curvatures of different signs are allowed, segments with negative curvature can be accounted for by generalizing $n(\boldsymbol{x}, \varphi, t)$ to a measure of the effective segment density, considering the continuously distributed generalization of

$$\frac{N_\varphi^+ - N_\varphi^-}{\Delta\Omega}, \tag{7.46}$$

where N_φ^+ and N_φ^- are the numbers of segments of orientation φ in $\Delta\Omega$ with positive and negative curvatures, respectively. In that case Eqns. (7.41) and (7.45) still hold. However, the (rather lengthy) derivation does not furnish new physical insights. The interested reader is referred to Hochrainer (2006), where the evolution equations of the density and the curvature are derived based on the higher-dimensional dislocation-density-tensor.

7.9 An Analytical Solution

To derive an analytical solution, an initial distribution of dislocation loops of homogeneous radius R_0 is considered. Let the distribution be independent of x_1 and increase linearly with respect to x_2

$$n(\boldsymbol{x}, \varphi, 0) = m_{n0}x_2 + b_{n0} + m_{n0}R_0\cos(\varphi). \tag{7.47}$$

The last term accounts for the constraint that dislocations must not end inside the crystal (see, e.g., Hochrainer, 2006). The evolution is governed by the dislocation velocity, which is prescribed to be constant $\nu = \nu_0$. Hence, all dislocation loops grow linearly in time

$$R_i = R_0 + \nu t \qquad \Rightarrow \quad k = \frac{1}{R_0 + \nu t}. \tag{7.48}$$

The segments of orientation φ move in the direction $\boldsymbol{e}_\nu(\varphi) = \sin(\varphi)\boldsymbol{e}_1 - \cos(\varphi)\boldsymbol{e}_2$. This motion preserves the independence of x_1 and yields the following solution

$$n = m_{n0}[x_2 + \nu_0 t\cos(\varphi)] + b_{n0} + m_{n0}R_0\cos(\varphi). \tag{7.49}$$

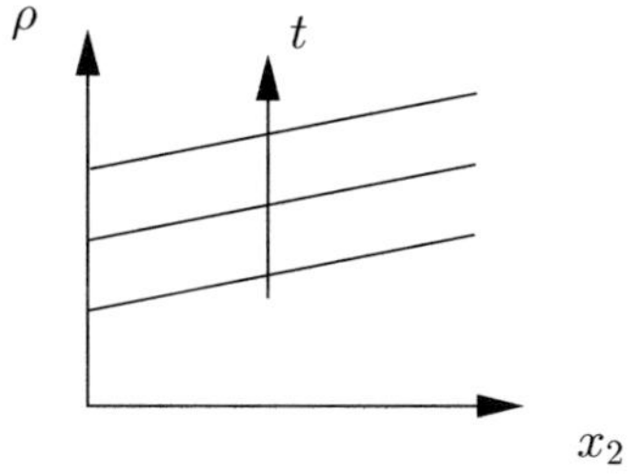

Figure 7.6: Qualitative representation of the analytical solution at different times.

The dislocation density is given by

$$\rho_\varphi = \frac{n}{k} = (R_0 + \nu_0 t)\left\{m_{n0}[x_2 + (R_0 + \nu_0 t)\cos(\varphi)] + b_{n0}\right\}. \tag{7.50}$$

These solutions satisfy the evolution equations (7.41) and (7.45).
The total dislocation density is given by

$$\rho = \oint \rho_\varphi \,\mathrm{d}\varphi = 2\pi(R_0 + \nu_0 t)(m_{n0}x_2 + b_{n0}). \tag{7.51}$$

It is visualized qualitatively in Fig. 7.6.

7.10 Averaged Equations

For practical applications, the solution of Eqns. (7.41) and (7.45) is, in general, numerically expensive due to the additional dimension φ. In order to eliminate the extra dimension, the system is integrated over the orientation space (Hochrainer et al., 2010; Hochrainer, 2013) and the problem is reformulated in terms of the total dislocation density

$$\rho = \oint \rho_\varphi \,\mathrm{d}\varphi, \tag{7.52}$$

curvature density

$$\overline{\rho k} = \oint \rho_\varphi k \,\mathrm{d}\varphi = \oint n \,\mathrm{d}\varphi \tag{7.53}$$

and dislocation density vector $\boldsymbol{\kappa}$

$$\boldsymbol{\kappa} = \oint \rho_\varphi \boldsymbol{e}_l \,\mathrm{d}\varphi. \tag{7.54}$$

The dislocation density vector $\boldsymbol{\kappa}$ can be expressed in terms of $\rho_{\vdash}$ and $\rho_{\odot}$ (see Eq. (1.24))

$$\boldsymbol{\kappa} = \frac{1}{b}(\rho_{\vdash}\boldsymbol{l} + \rho_{\odot}\boldsymbol{d}), \tag{7.55}$$

i.e., it is a GND measure. Here, it has been assumed that the initial plastic shear $\gamma|_{t=0}$ vanishes.
For simplicity, the special case of an isotropic velocity is considered, i.e., $\nu = \nu(\boldsymbol{x}, t)$. In this case, the angular velocity is given by $\vartheta(\boldsymbol{x}, \varphi, t) = -\boldsymbol{e}_{\mathrm{l}}(\varphi) \cdot \nabla\nu(\boldsymbol{x}, t)$.
The vector valued average dislocation velocity is given by

$$\frac{1}{\rho}\oint \rho_{\varphi}\boldsymbol{\nu}\,\mathrm{d}\varphi = \frac{\boldsymbol{\kappa}^{\perp}\nu}{\rho} = \left(\frac{\kappa}{\rho}\nu\right)\boldsymbol{e}_{\kappa^{\perp}}, \tag{7.56}$$

with $\kappa = \|\boldsymbol{\kappa}\|$ and $\boldsymbol{\kappa}^{\perp} = -\nabla_{\mathrm{p}}\gamma/b = -(\boldsymbol{I} - \boldsymbol{n}\otimes\boldsymbol{n})\nabla\gamma/b$.
Integration of Eqns. (7.41) and (7.45) yields

$$\begin{aligned} &\partial_t\rho + \operatorname{div}\left(\boldsymbol{\kappa}^{\perp}\nu\right) = \overline{\rho k}\nu, \\ &\partial_t\overline{\rho k} + \operatorname{div}\left(\oint \rho_{\varphi}k\boldsymbol{e}_{\nu}\,\mathrm{d}\varphi\nu + \oint \rho\boldsymbol{e}_{\mathrm{l}}\otimes\boldsymbol{e}_{\mathrm{l}}\,\mathrm{d}\varphi\nabla\nu\right) = 0. \end{aligned}$$

Further simplifying assumptions are required to evaluate the remaining integrals. The integrals can be simplified, if the curvature k (first integral) and density ρ_{φ} (second integral), respectively, are assumed independent of the orientation (Hochrainer et al., 2010)

$$\partial_t\rho + \operatorname{div}\left(\boldsymbol{\kappa}^{\perp}\nu\right) = \overline{\rho k}\nu, \tag{7.57}$$

$$\partial_t\overline{\rho k} + \operatorname{div}\left(\bar{k}\boldsymbol{\kappa}^{\perp}\nu + \frac{1}{2}\rho\nabla_{\mathrm{p}}\nu\right) = 0, \tag{7.58}$$

with $\bar{k} = \overline{\rho k}/\rho$ and $\nabla_{\mathrm{p}}\nu = (\boldsymbol{I} - \boldsymbol{n}\otimes\boldsymbol{n})\nabla\nu$. An alternative approach of the simplification of the two integrals can be found in Hochrainer et al. (2014).

Chapter 8

A Simplified Thermodynamic Continuum Dislocation Dynamics Framework – Theoretical Aspects and Finite Element Simulation

8.1 Introduction

In order to enhance the gradient plasticity theory developed in Chaps. 4 and 5, the kinematical framework of the model is subsequently extended. A significantly more detailed description of the dislocation microstructure is established by introducing two additional field variables per slip system: the total dislocation density ρ and the average dislocation curvature $\bar{k}$, as defined in Chap. 7. These quantities extend the kinematical setting of Chaps. 4 and 5, which extended a classical crystal plasticity framework by the gradient of the equivalent plastic strain.

The evolution of the total dislocation densities and curvatures of the different slip systems is governed by the partial differential equations (7.57) and (7.58), which were derived by Hochrainer et al. (2010). Orowan's equation (Orowan, 1934) kinematically couples the additional dislocation fields with the plastic slip rates.

The extended kinematical basis can then be exploited to improve the constitutive model. In this regard, the principal aim of this chapter represents the application of a dislocation forest hardening model, which allows for the determination of the local yield stress as a function of the dislocation densities.

Although the improvement of the forest hardening model is the major aim, additional modifications of the model might be useful. For example, a more realistic form of the stored energy density function may be possible by introducing the dislocation densities and curvatures as additional arguments of the function. Unfortunately, the associated optimal form of the energy density function is (in general) unknown.

Nevertheless, it is worthwhile to investigate the system equations associated to such a modification. The introduction of new energy function arguments implies additional conjugate forces. The latter are accounted for when the principle of virtual power is evaluated, yielding an extended set of balance equations. Here, the application of the principle of virtual power differs slightly from its exploitation in previous chapters, since the evolution equations (7.57) and (7.58) of the additional field variables contain transport terms. The transport terms distinguish the model at hand from many other extended field theories like, for example, usual gradient extended standard dissipative solids and other Cosserat-like models.

An exemplary simple energy ansatz is investigated based on the assumption that dislocation interactions can be neglected, i.e., an extra energy term is introduced which is proportional to the total dislocation density. This ansatz is chosen for illustration reasons, i.e., its physical relevance remains to be discussed. Here, it is applied since it implies an additional backstress term in the flow rule related to the dislocation line tension and curvature.

Finally, a numerical example illustrates the model behavior in a single slip situation, where the curvature dependent backstress term is neglected. In particular, the transport properties of the model are highlighted. The content of this chapter has partially been taken from Wulfinghoff and Böhlke (2012b).

8.2 Kinematical Framework

8.2.1 Distorsion and Dislocation Measures

In line with the previous chapters, the geometrically linear theory is applied. Accordingly, the displacement gradient and the strain tensor are assumed to be decomposed additively into elastic and plastic parts (Eqns. (1.10) and (1.11))

$$\boldsymbol{H} = \boldsymbol{H}^{\mathrm{e}} + \boldsymbol{H}^{\mathrm{p}}, \quad \boldsymbol{\varepsilon} = \boldsymbol{\varepsilon}^{\mathrm{e}} + \boldsymbol{\varepsilon}^{\mathrm{p}}. \tag{8.1}$$

Here, the plastic part of the deformation gradient takes the form

$$\boldsymbol{H}^{\mathrm{p}} = \sum_{\alpha=1}^{N} \lambda_\alpha \boldsymbol{d}_\alpha \otimes \boldsymbol{n}_\alpha. \tag{8.2}$$

Moreover, the equivalent plastic strain reads (Eq. (4.87))

$$\gamma_{\mathrm{eq}} = \sum_{\alpha} \int \dot{\lambda}_\alpha \, \mathrm{d}t. \tag{8.3}$$

The screw- and edge-components of the geometrically necessary dislocations (Eq. (1.24)) are interpreted as the components of the dislocation density vectors $\boldsymbol{\kappa}_\alpha$ (times the absolute values of the corresponding Burgers vectors, Eq. (7.55))

$$\boldsymbol{\alpha} = \mathrm{curl}^{\mathrm{T}}(\boldsymbol{H}^{\mathrm{p}}) = \sum_{\alpha=1}^{N} \boldsymbol{d}_\alpha \otimes (\rho_{\vdash}^{\alpha} \boldsymbol{l}_\alpha + \rho_{\odot}^{\alpha} \boldsymbol{d}_\alpha) = \sum_{\alpha=1}^{N} \boldsymbol{b}_\alpha \otimes \boldsymbol{\kappa}_\alpha, \tag{8.4}$$

with $\boldsymbol{l}_\alpha = \boldsymbol{n}_\alpha \times \boldsymbol{d}_\alpha$, $\rho_{\vdash}^{\alpha} = -\boldsymbol{d}_\alpha \cdot \nabla \gamma_\alpha$ and $\rho_{\odot}^{\alpha} = \boldsymbol{l}_\alpha \cdot \nabla \gamma_\alpha$ (Eq. (1.24)).
In addition to the aforementioned GND-measures, the dislocation microstructure is represented by the total dislocation densities ρ_α and average curvatures $\bar{k}_\alpha$ of the individual slip systems (see Sect. 7.10). Here, ρ_α is interpreted as the "total line length per unit volume".

8.2.2 Kinematical Equations

The dislocation fields ρ_α and $\bar{k}_\alpha$ evolve according to Eqns. (7.57) and (7.58) which were derived by Hochrainer et al. (2010)

$$\begin{aligned} \partial_t \rho_\alpha &= -\mathrm{div}\left(\nu_\alpha \boldsymbol{\kappa}_\alpha^\perp\right) + \overline{\rho_\alpha k_\alpha} \nu_\alpha, & (8.5)\\ \partial_t \overline{\rho_\alpha k_\alpha} &= -\mathrm{div}\left(\bar{k}_\alpha \nu_\alpha \boldsymbol{\kappa}_\alpha^\perp\right) - \frac{1}{2}\mathrm{div}\left(\rho_\alpha \nabla_\mathrm{p} \nu_\alpha\right), & (8.6) \end{aligned}$$

with $\overline{\rho_\alpha k_\alpha} = \rho_\alpha \bar{k}_\alpha$ and $\boldsymbol{\kappa}_\alpha^\perp = -(\boldsymbol{I} - \boldsymbol{n}_\alpha \otimes \boldsymbol{n}_\alpha)\nabla\gamma_\alpha / b_\alpha$. As mentioned in Sect. 7.10, the dislocation velocity ν_α is assumed to be independent of the dislocation orientation.

Equations (8.5) and (8.6) are purely kinematical evolution equations that account for curvature-induced dislocation line production and dislocation transport. They require proper initial and boundary conditions. The production of new lines (for example as a consequence of Frank-Read sources) is not included.

The dislocation fields ρ_α and ν_α are kinematically coupled to the plastic slip rates by Orowan's equation (1.4)

$$\dot{\gamma}_\alpha = \rho_\alpha b_\alpha \nu_\alpha. \qquad (8.7)$$

The weak forms of Eqns. (8.5) and (8.6) are given by

$$\int_\mathcal{B} \left((\partial_t \rho_\alpha - \overline{\rho_\alpha k_\alpha} \nu_\alpha) w_{\rho,\alpha} - \nu_\alpha \boldsymbol{\kappa}_\alpha^\perp \cdot \nabla w_{\rho,\alpha}\right) \mathrm{d}v + \int_{\partial\mathcal{B}} w_{\rho,\alpha} \nu_\alpha \boldsymbol{n} \cdot \boldsymbol{\kappa}_\alpha^\perp \, \mathrm{d}a = 0 \quad (8.8)$$

and

$$\begin{aligned} & \int_\mathcal{B} \left(\partial_t \overline{\rho_\alpha k_\alpha} w_{\rho k,\alpha} - (\nu_\alpha \bar{k}_\alpha \boldsymbol{\kappa}_\alpha^\perp + (1/2)\rho_\alpha \nabla_\mathrm{p} \nu_\alpha) \cdot \nabla w_{\rho k,\alpha}\right) \mathrm{d}v \\ + & \int_{\partial\mathcal{B}} w_{\rho k,\alpha} (\nu_\alpha \bar{k}_\alpha \boldsymbol{\kappa}_\alpha^\perp + (1/2)\rho_\alpha \nabla_\mathrm{p} \nu_\alpha) \cdot \boldsymbol{n} \, \mathrm{d}a = 0. \qquad (8.9) \end{aligned}$$

Here, $w_{\rho,\alpha}$ and $w_{\rho k,\alpha}$ denote arbitrary test functions.

8.3 Principle of Virtual Power

In order to derive field and boundary conditions for a single crystal, the principle of virtual power (Chap. 4, Eq. (4.14)) is exploited. Based on the extended kinematical framework outlined in Sect. 8.2, additional generalized stresses are introduced conjugate to $\dot{\zeta}$, $\nabla\dot{\zeta}$, $\dot{\gamma}_\alpha$ and $\dot{\rho}_\alpha$. It turns out that an advantageous choice of the independent kinematical variables is given by the displacement field $\boldsymbol{u}$, the plastic variable ζ and the plastic slip parameters λ_α. Like in Chaps. 4 and 5, two slip parameters are introduced for each crystallographic slip system. These account for the positive and negative slip direction, respectively. Let the slip parameters with indices 1 to N be associated to the positive slip directions (Eq. (4.78)).

The plastic slips γ_α are dependent variables and can be obtained from the differences of the associated slip parameters (Eq. (4.85)). The virtual quantities are assumed to respect this dependency, i.e. $\delta\dot{\gamma}_\alpha = \delta\dot{\lambda}_\alpha - \delta\dot{\lambda}'_\alpha$. In addition, the virtual slip rates are chosen to respect Orowan's equation (8.7)

$$\delta\dot{\gamma}_\alpha = \rho_\alpha b_\alpha \delta\nu_\alpha, \tag{8.10}$$

where $\delta\nu_\alpha$ denote the virtual dislocation velocities. Moreover, the virtual rates of change of the dislocation densities are given by (see Eq. (8.5))

$$\delta\dot{\rho}_\alpha = -\operatorname{div}\left(\delta\nu_\alpha \boldsymbol{\kappa}_\alpha^\perp\right) + \overline{\rho_\alpha k_\alpha}\delta\nu_\alpha. \tag{8.11}$$

As already mentioned, the principle of virtual power (4.14) is extended. The virtual power of internal forces is postulated by

$$\delta\mathcal{P}_{\text{int}} = \int_{\mathcal{B}} \left(\boldsymbol{\sigma}\cdot\delta\dot{\boldsymbol{\varepsilon}} + \xi\,\delta\dot{\zeta} + \boldsymbol{\xi}\cdot\nabla\delta\dot{\zeta} + \sum_\alpha(\tau_\alpha^{\text{k}}\delta\dot{\gamma}_\alpha + \eta_\alpha\delta\dot{\rho}_\alpha)\right)\mathrm{d}v. \tag{8.12}$$

Here, τ_α^{k} and η_α are generalized stresses conjugate to $\dot{\gamma}_\alpha$ and $\dot{\rho}_\alpha$.

Virtual power contributions associated to the curvatures $\bar{k}_\alpha$ are neglected, for simplicity. Moreover, grain boundary contributions do not arise since the body under consideration is a single crystal.

In analogy to the virtual power of internal forces, the external part is extended, too

$$\delta\mathcal{P}_{\text{ext}} = \int_{\partial\mathcal{B}_t} \bar{\boldsymbol{t}}\cdot\delta\dot{\boldsymbol{u}}\,\mathrm{d}a + \int_{\partial\mathcal{B}_\Xi} \bar{\Xi}\,\delta\dot{\zeta}\,\mathrm{d}a + \sum_\alpha \int_{\partial\mathcal{B}_{t\alpha}} \bar{t}_\alpha\,\delta\nu_\alpha\,\mathrm{d}a, \tag{8.13}$$

where the generalized tractions $\bar{t}_\alpha$ arise in addition to the tractions $\bar{\boldsymbol{t}}$ and microtractions $\bar{\Xi}$. They are assumed to be defined at the boundaries $\partial\mathcal{B}_t$, $\partial\mathcal{B}_\Xi$ and $\partial\mathcal{B}_{t_\alpha}$, respectively. In addition, Dirichlet boundary conditions are considered at the remaining parts of the boundary. In this work, exclusively microhard Dirichlet boundary conditions are considered for the plastic variables λ_α and ζ. These are regions of the boundary which are modeled impenetrable for dislocations, i.e., all plastic field variables are assumed to vanish. This implies equality of the remaining parts $\partial\mathcal{B}_\Xi = \partial\mathcal{B}_{t_\alpha}$, $\alpha = 1, \ldots, N$.
With the same arguments as in Sect. 4.3 (application of Gauss' theorem and the chain rule) as well as Eqns. (8.10) and (8.11), the following field equations follow from the principle of virtual power $\delta\mathcal{P}_{\text{int}} = \delta\mathcal{P}_{\text{ext}}$

$$\operatorname{div}(\boldsymbol{\sigma}) = \mathbf{0}, \tag{8.14}$$

$$\operatorname{div}(\boldsymbol{\xi}) = \xi, \tag{8.15}$$

$$\frac{1}{\rho_\alpha b_\alpha^2}\nabla\eta_\alpha \cdot \nabla_{\text{p}}\gamma_\alpha - \frac{\bar{k}_\alpha\eta_\alpha}{b_\alpha} = \tau_\alpha^{\text{k}}, \quad \alpha = 1, \ldots, N. \tag{8.16}$$

This set of equations represents an extended version of Eqns. $(4.16)_1$ and $(4.17)_1$ derived in Chap. 4.3. In addition, the following boundary equations arise (compare Eqns. $(4.16)_2$ and $(4.17)_2$)

$$\bar{\boldsymbol{t}} = \boldsymbol{\sigma}\boldsymbol{n} \quad \text{on } \partial\mathcal{B}_t, \tag{8.17}$$

$$\bar{\Xi} = \boldsymbol{\xi}\cdot\boldsymbol{n} \quad \text{on } \partial\mathcal{B}_\Xi, \tag{8.18}$$

$$\bar{t}_\alpha = \boldsymbol{n}\cdot\nabla_{\text{p}}\gamma_\alpha\frac{\eta_\alpha}{b_\alpha} \quad \text{on } \partial\mathcal{B}_{t_\alpha}, \quad \alpha = 1, \ldots, N. \tag{8.19}$$

Eqns. (8.15) and (8.16) are generalized force balance equations, which extend the linear momentum balance (8.14). They are direct consequences of the introduction of the generalized forces, which are conjugate to the additional dislocation field variables.

8.4 Energy Density and Dissipation

In a quite general setting, the stored energy density may be assumed to be a function of the different state and history variables introduced in

Sect. 8.2

$$W = W(\boldsymbol{\varepsilon}, \boldsymbol{\varepsilon}^{\mathrm{p}}, \hat{\rho}, \hat{k}, \hat{\rho}_{\vdash}, \hat{\rho}_{\odot}, \dots), \tag{8.20}$$

where the following arrays have been introduced

$$\hat{\rho} = (\rho_1, \rho_2, \dots, \rho_N)^{\mathrm{T}}, \quad \hat{k} = (\bar{k}_1, \bar{k}_2, \dots, \bar{k}_N)^{\mathrm{T}} \tag{8.21}$$

and

$$\hat{\rho}_{\vdash} = (-\boldsymbol{d}_1 \cdot \nabla\gamma_1, -\boldsymbol{d}_2 \cdot \nabla\gamma_2, \dots, -\boldsymbol{d}_N \cdot \nabla\gamma_N)^{\mathrm{T}}, \tag{8.22}$$

$$\hat{\rho}_{\odot} = (\boldsymbol{l}_1 \cdot \nabla\gamma_1, \boldsymbol{l}_2 \cdot \nabla\gamma_2, \dots, \boldsymbol{l}_N \cdot \nabla\gamma_N)^{\mathrm{T}}. \tag{8.23}$$

In this work, the general approach (8.20) is simplified as follows.

- In order to reduce the computational effort, the arrays of projected slip gradients $\hat{\rho}_{\vdash}$ and $\hat{\rho}_{\odot}$ are replaced by the gradient of the plastic field variable $\nabla\zeta$.
- In addition, the influence of the curvatures $\hat{k}$ on the energy density function is neglected.

As a result, the energy density function can be represented as follows

$$W = W(\boldsymbol{\varepsilon}, \boldsymbol{\varepsilon}^{\mathrm{p}}, \hat{\rho}, \nabla\zeta). \tag{8.24}$$

For simplicity, it is assumed that the following stresses are energetic

$$\boldsymbol{\sigma} = \partial_{\boldsymbol{\varepsilon}} W, \quad \boldsymbol{\xi} = \partial_{\nabla\zeta} W, \quad \eta_\alpha = \partial_{\rho_\alpha} W. \tag{8.25}$$

It can be shown that in this case, the reduced dissipation inequality reads

$$\mathcal{D} = \xi\dot{\zeta} + \sum_{\alpha}(\tau_\alpha + \tau_\alpha^{\mathrm{k}})\dot{\lambda}_\alpha \geq 0. \tag{8.26}$$

Here, the same arguments as in Sect. 4.4.1 (Eq. (4.27)) have been applied. As has been mentioned earlier, each slip system is described by two slip parameters λ_α. Note that the associated resolved shear stresses τ_α have the same magnitude but different signs. The same holds for the stresses τ_α^{k}.

Example: The mesoscopic elastic energy is assumed to be given by $W_{\mathrm{e}} = \boldsymbol{\varepsilon}^{\mathrm{e}} \cdot \mathbb{C}[\boldsymbol{\varepsilon}^{\mathrm{e}}]/2$. Any energetic interaction of the different arguments $\boldsymbol{\varepsilon}^{\mathrm{e}}$, $\hat{\rho}$ and $\nabla\zeta$ is neglected, i.e.,

$$W = W_{\mathrm{e}}(\boldsymbol{\varepsilon}, \boldsymbol{\varepsilon}^{\mathrm{p}}) + W_{\rho}(\hat{\rho}) + W_{\mathrm{g}}(\nabla\zeta). \tag{8.27}$$

This is a strong simplification. Nevertheless, the identification of suitable forms of the individual contributions $W_{\mathrm{e}}(\boldsymbol{\varepsilon}, \boldsymbol{\varepsilon}^{\mathrm{p}})$, $W_{\rho}(\hat{\rho})$ and $W_{\mathrm{g}}(\nabla\zeta)$ remains a non-trivial task, which is not within the scope of the work at hand. Instead, further simplified assumptions are made for W_{ρ} and W_{g}. For example, a simple approach may be

$$W_{\rho}(\hat{\rho}) = c_{\mathrm{L}} \sum_{\alpha} \rho_{\alpha}, \quad W_{\mathrm{g}}(\nabla\zeta) = \frac{1}{2} E l^2 \nabla\zeta \cdot \nabla\zeta, \tag{8.28}$$

where c_{L} can be interpreted as dislocation self-energy per unit length and l is an internal length scale parameter.
As a consequence, the dissipation (8.26) takes the form

$$\mathcal{D} = \dot{\zeta}\, E l^2 \Delta\zeta + \sum_{\alpha} (\tau_{\alpha} \pm c_{\mathrm{L}} \bar{k}_{\alpha}/b_{\alpha}) \dot{\lambda}_{\alpha} \geq 0. \tag{8.29}$$

Here, Eqns. (8.15) and (8.16) have been exploited. Obviously, the energy contribution $W_{\rho}(\hat{\rho})$ (Eq. $(8.28)_1$) implies a line tension induced backstress which is proportional to the average curvature. The sign of the backstress depends on whether the associated slip parameter refers to the positive or negative slip direction (where it has been assumed that $k_{\alpha+N} = k_{\alpha}$).

8.5 Yield Criteria and Flow Rule

Motivated by Eq. (8.26), the following yield criteria are proposed

$$f_{\alpha} = \tau_{\alpha}^{\mathrm{eff}} + \xi - \tau_{\alpha}^{\mathrm{C}}, \tag{8.30}$$

where $\tau_{\alpha}^{\mathrm{eff}} = \tau_{\alpha} + \tau_{\alpha}^{\mathrm{k}}$ is an effective shear stress and $\tau_{\alpha}^{\mathrm{C}} = \tau_{\alpha}^{\mathrm{C}}(\hat{\rho})$ denotes an offset yield stress due to forest hardening and lattice friction.

In addition, thermodynamically consistent viscoplastic flow rules are given by

$$\dot{\lambda}_\alpha = \Lambda_\alpha \partial_{\tau_\alpha^{\mathrm{eff}}} f_\alpha = \Lambda_\alpha, \tag{8.31}$$

$$\dot{\zeta} = \sum_\alpha \Lambda_\alpha \partial_\xi f_\alpha = \sum_\alpha \Lambda_\alpha. \tag{8.32}$$

Here, $\Lambda_\alpha = \Lambda_\alpha(f_\alpha)$ are overstress type functions satisfying

$$\Lambda_\alpha(f_\alpha) \geq 0, \tag{8.33}$$

$$\Lambda_\alpha(f_\alpha) = 0, \text{ if } f_\alpha \leq 0. \tag{8.34}$$

Like in Chap. 4 (Eq.(4.88)), the field variable ζ can be identified by the equivalent plastic strain

$$\zeta = \gamma_{\mathrm{eq}}. \tag{8.35}$$

Example: The previous example (based on Eqns. (8.27) and (8.28)) is continued. A possible choice of the overstress functions $\Lambda_\alpha(f_\alpha)$ is the power law

$$\Lambda_\alpha = \rho_\alpha b_\alpha \nu_0 \left\langle \frac{f_\alpha}{\tau^{\mathrm{D}}} \right\rangle^p = \rho_\alpha b_\alpha \nu_0 \left\langle \frac{\tau_\alpha \pm (c_{\mathrm{L}} \bar{k}_\alpha / b_\alpha) + E l^2 \Delta\gamma_{\mathrm{eq}} - \tau_\alpha^{\mathrm{C}}(\hat{\rho})}{\tau^{\mathrm{D}}} \right\rangle^p, \tag{8.36}$$

where ν_0 is a dislocation reference velocity and the remaining material parameters have the same interpretation as in the previous chapters. In Eq. (8.36), Eqns. (8.15), (8.16) and (8.30) have been exploited.
Compared to the flow rule (4.83) of the gradient plasticity model introduced in Sect. 4.6, the flow rule (8.36) additionally includes a line tension induced backstress. Moreover, the yield stress $\tau_\alpha^{\mathrm{C}}(\hat{\rho})$ is modeled as a function of the dislocation densities. This model replaces the phenomenological hardening stress $\beta^{\mathrm{e}} = \partial_\zeta W$ of flow rule (4.83).

8.6 Numerical Example

The model is discretized in space by the Finite Element Method. For simplicity, a single slip situation is considered and the slip system index

is dropped. Moreover, the elastic properties are assumed isotropic and described by Young's modulus E and Poisson's ratio υ.
The following approaches are applied

$$W_{\rm g}(\nabla\gamma_{\rm eq}) = \frac{1}{2}K_{\rm G}\nabla\gamma_{\rm eq}\cdot\nabla\gamma_{\rm eq}, \quad \tau^{\rm C} = \tau_0^{\rm C} + a_{\rm T}Gb\sqrt{\rho}, \tag{8.37}$$

where G is the shear modulus, $\tau_0^{\rm C}$ denotes the lattice friction stress and $a_{\rm T}$ is a constant of order unity (see Eq. (1.21)). The gradient modulus $K_{\rm G} = El^2$ implicitly defines an internal length scale parameter l.

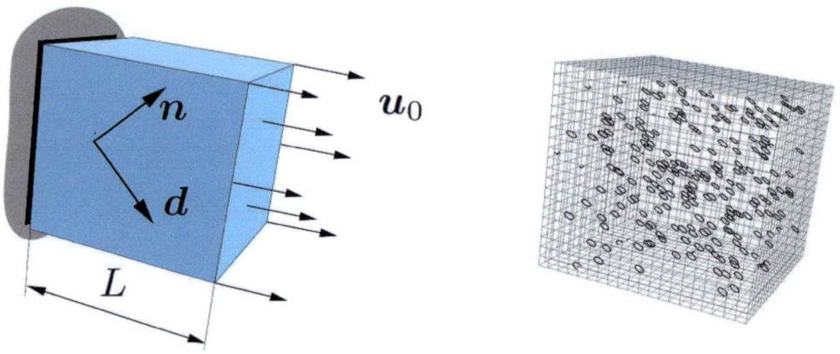

Figure 8.1: Simulation set up.

The term $W_\rho(\rho) = c_{\rm L}\rho$ is not considered for three reasons.

1. The physical relevance of the ansatz $W_\rho(\rho) = c_{\rm L}\rho$ is not verified. It has been chosen for illustration reasons. Moreover, it implies a backstress term (Eq. (8.36)) which is associated to the line tension and curvature. This is illustrative, but not necessarily physically correct.
2. The influence of the curvature $\bar{k}$ is negligible if it is sufficiently large. It may be assumed that a reasonable value of $c_{\rm L}$ should be equal to or less than the dislocation self-energy per unit length of a single dislocation. The latter is given by $\sim 1\,{\rm nJ/m}$, i. e., the value of the backstress $c_{\rm L}\bar{k}/b$ is expected to be less than $\sim 0.01{\rm MPa}$ for $\bar{k} = 1\mu{\rm m}^{-1}$ (which is the order of magnitude observed in the simulations) and therefore negligible.

3. Yet, the effects of the curvature-induced backstress on the numerical implementation have not been studied and might be of complicated type (e. g. softening phenomena, localization).

A linear viscous model is applied, i.e., the rate sensitivity parameter p of the flow rule (8.36) is set to one. The parameters ν_0 and τ^{D} are chosen such that the viscosity-like quantity τ^{D}/ν_0 is small, i.e., the results are close to the rate-independent case.
In detail, the material parameters are given by

K_{G}	E	υ	τ_0^{C}	a_{T}	ν_0	τ^{D}
0.1 N	70 GPa	0.34	20MPa	0.2	1m/s	1 MPa

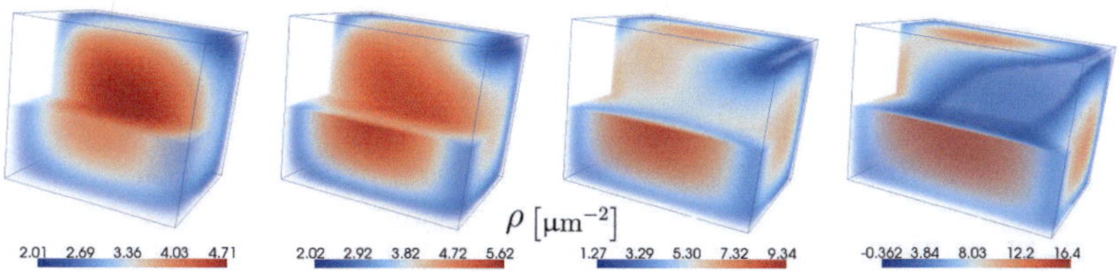

Figure 8.2: Evolution of the dislocation density ρ for 30%, 40%, 60% and 100% of the deformation. One quarter of the body has been removed to show the bulk response. The deformation of the body is scaled by a factor of 50.

A tensile test of a cube-shaped crystal is simulated. The edge length is given by $L = 10\,\mu\mathrm{m}$. Micro-hard boundary conditions and prescribed displacements $\boldsymbol{u}_0 = u_0 t/T \boldsymbol{e}_1$ are applied on one face (cf. Fig. 8.1). The total simulation time and final displacements are given by $T = 1\,\mathrm{s}$ and $u_0 = 0.05\,\mu\mathrm{m}$, respectively. Accordingly, the ultimate average tensile strain is 0.5%. Furthermore, the displacement boundary conditions allow for free lateral contractions.
The body is discretized by $20 \times 20 \times 20$ standard linear hexahedrons. For simplicity, dislocation boundary flux terms are assumed to vanish. The slip system is rotated about 45° with respect to the tensile direction (see Fig. 8.1).

The initial conditions of the dislocation fields are given by $\rho(\boldsymbol{x}, t = 0) = 2\mu\text{m}^{-2}$ and $\overline{\rho k}(\boldsymbol{x}, t = 0) = 2\mu\text{m}^{-3}$. Accordingly, the initial curvature is $\bar{k}(\boldsymbol{x}, t = 0) = 1\mu\text{m}^{-1}$. Physically, these initial conditions correspond to a homogeneous distribution of small loops (cf. Fig. 8.1).
The problem is solved in 100 equal time steps. In each step, the following quantities are implicitly solved for

- the displacement increments (as nodal unknowns)
- the increments of the equivalent plastic strain (as nodal unknowns)
- the plastic slip parameter increments (at the integration points)

The dislocation velocities are assumed constant during the individual time steps. They are computed at the integration points based on Orowan's equation

$$\Delta\gamma = \rho_{\text{n}} b \nu \Delta t. \tag{8.38}$$

Having identified the dislocation velocities at the integration points, those are mapped to the nodes by a L2-projection (based on a diagonalized mass matrix, see e. g. Becker (2006)). The weak forms (8.8) and (8.9) of the dislocation evolution equations (8.5) and (8.6) are subsequently solved based on the explicit Euler scheme (again based on a diagonalized mass matrix). Then, the next time step can be solved. To prevent the algorithm from producing large negative density values due to numerical errors, the viscosity is set to a high value in regions where the dislocation density ρ falls below a critical threshold (i.e. a plastic deformation is not possible if no dislocations are present).
In the following, the simulation results are interpreted. Due to the prescribed displacements, the resolved shear stress increases until it reaches the yield stress τ^{C}. Subsequently, the dislocation motion is triggered, i.e., the continuously distributed loops start to grow (curvature-induced line length production). The dislocation density "piles up" at the boundaries (see Fig. 8.3) and, as a consequence, slows down the density production close to the boundaries (cf. Fig 8.2 at 30% of deformation). At the early stage of the deformation, the dislocation

microstructure is SSD-dominated. This result is reasonable because, at this point of the deformation process, the orientations of the dislocation lines (which are represented by ρ) in the bulk can be considered nearly uniformly distributed (overlapping growing circles).

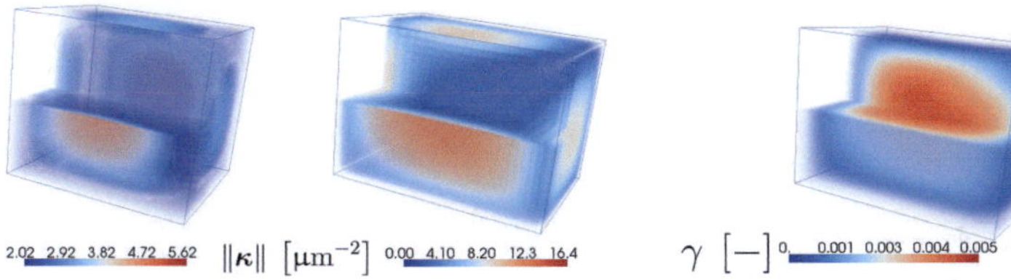

Figure 8.3: The geometrically necessary dislocation density $\|\kappa\|$ for 40% and 100% of the deformation and the final plastic slip γ.

As the deformation advances, the dislocation lines are expected to grow until they become pile-ups at the micro-hard boundary. Indeed, the simulation shows that the dislocation density, which was produced in the bulk, is transported towards the boundary (Fig. 8.2). The evaluation of the geometrically necessary density $\|\kappa\|$ shows that

1. at the final state nearly the total dislocation density is geometrically necessary (cf. Figs. 8.2 and 8.3 at 100% deformation),
2. the predictions of the kinematical theory, given by Eqns. (8.5) and (8.6) is consistent with the GND-density predictions which are computed from the gradients of the plastic slips.

The latter point verifies that the dislocation evolution Eqns. (8.5) and (8.6) represent the dislocation transport and production in a reasonable way.

Chapter 9

Summary and Outlook

This thesis addresses various open questions in gradient and micro plasticity. Amongst other things, the following aspects of gradient crystal plasticity are considered

- the form of the defect energy density function
- a possible illustration of the generalized stresses and microtractions
- the decreased numerical efficiency due to the need for additional nodal degrees of freedom and a simplified, numerically efficient and robust gradient theory including a grain boundary model
- the capability of grain boundary models to capture the Hall-Petch effect
- dislocation density based work-hardening models taking into account dislocation transport

The first point is related to the defect energy density function, which is a key ingredient of most gradient plasticity theories. Besides the quadratic approach, several non-smooth energies have been proposed in the literature. The work at hand describes these by sub-differentials, which are also applied in the rate-variational description of local plasticity models.

A non-smooth rank-one energy is the first one of two promising defect energy candidates that are investigated in the work at hand. This type of potential has been used by several authors and is known to lead to a size-dependent overall yield stress in certain situations. The second candidate is a regularized logarithmic energy function, leading

to a gradient plasticity approximation of Groma's theory. Illustrative physical interpretations of all material parameters are given for both energy functions.
The features of both energies are investigated through an exemplary application to a laminate model material. In both cases, a size-dependent overall yield stress and Asaro's type KIII hardening are obtained.
Subsequently, it is shown that the model is fully represented by a mixed incremental potential including Lagrange multipliers which are identified with the generalized stresses. The potential is used as starting point for the finite element discretization introducing the displacements and the plastic distorsion as nodal degrees of freedom. In three-dimensional applications, this implies twelve nodal unknowns.
The incremental potential structure, including the aforementioned Lagrange multipliers, facilitates a new numerical treatment of non-smooth defect energies like, e.g., the rank-one potential. The sub-differential description reveals an analogy to rigid-plasticity, which may be regularized by introducing small elastic deformations. This type of regularization is known to yield robust and efficient numerical algorithms. In the work at hand, this regularization technique is applied to non-smooth defect energies and further developed in the form of an augmented Lagrange multiplier algorithm.
The gradient plasticity framework used to solve the laminate problem exploits the principle of virtual power, which requires the introduction of generalized stresses. These are, e.g., conjugate to the dislocation density tensor and closely related to the microtractions at the boundary. In the subsequent investigation, these rather non-intuitive quantities are exemplarily illustrated for a special model material, which is characterized by a highly correlated dislocation microstructure. More precisely, the dislocations are assumed to be arranged in ideally localized slip bands. The regions in between the slip bands are assumed to deform purely elastically. It is argued that, under certain conditions, the material model may be approximated by a gradient plasticity theory.
This approximation facilitates an illustrative interpretation of several

usually non-intuitive features of the gradient theory. For example, the generalized stresses can be identified with microscopic Cauchy stress fluctuations in elastic regions in between the slip bands. The internal length scale is found to be given by the slip band distance. Moreover, the microscopic stress fluctuations are similar to stress distributions known from bending and torsion theories. In addition, the defect energy coefficients can be interpreted as generalized, distributed moments of inertia. The resultant moments of the stress fluctuations may be related to the microtractions of the gradient model. For simple systems, the microforce balance is related to the differential moment equations known from bending and torsion theory.

Having obtained this illustrative interpretation of the generalized stresses for the considered model material, the subsequent chapter returns to a more abstract description. The principal idea is to formulate a simplified, numerically efficient theory, being particularly suited for three-dimensional multislip and multiphase applications. Therefore, the independent kinematical fields are chosen to be the displacement field and a phenomenological scalar plastic field variable. This implies a significant reduction of the number of nodal degrees of freedom compared to most gradient crystal plasticity theories.

The exploitation of the principle of virtual power yields the balance equations. In this regard, grain boundaries are explicitly accounted for through a respective microtraction balance equation.

The system of equations is closed by constitutive assumptions on, e.g., the energy density and the yield criteria. In this context, the principle of maximum dissipation is exploited in order to derive thermodynamically consistent flow rules for the rate-independent case. In addition, viscoplastic flow rules of the overstress type are discussed.

Although thermodynamically consistent, the theory is not free of conceptual difficulties. These arise due to the inability of the scalar plastic field variable to carry any information on the plastic flow direction. As a consequence, the solution may become unstable with respect to infinitesimal changes of the boundary conditions. This problem can

partially be resolved by the definition of separate yield criteria for opposed flow directions. A gradient crystal plasticity framework is presented adopting this concept.
Subsequently, a micromorphic reformulation of this model is outlined. The micromorphic framework facilitates an efficient finite element implementation and integration point wise evaluation of the flow rule. In particular, a numerically robust algorithm for the grain boundary treatment is discussed. Moreover, an enhanced power law subroutine is presented which can handle large time steps, even if the flow rule is strongly nonlinear due to a large strain rate sensitivity.
The model parameters are fitted to experimental tensile test data of copper microwires with variable grain size. The data indicates that the grain size effect mainly affects the overall yield strength of the wires. In contrast, the overall hardening barely shows any size effect, i.e., the wires exhibit the typical Hall-Petch behavior.
The proposed model is capable to capture this experimentally observed size effect. For the investigated load case and parameters, the size effect of the model is mainly controlled by the grain boundary resistance. In contrast, the defect energy material parameters could be varied within a wide range without having a strong effect on the results. It is expected that the defect energy has a stronger influence on the results if more heterogeneous deformation modes than tension are considered.
The subsequent study investigates the theory's general potential to reproduce the Hall-Petch scaling law. A rough estimation suggests that the overall yield strength rather scales like $1/D$ than $1/\sqrt{D}$, where D is the average grain size. In order to better understand this deficiency, the theory is compared to a model which is close to the one proposed by Eshelby et al. (1951). This comparison is motivated by the fact that the model of Eshelby et al. (1951) represents a wide-spread explanation of the Hall-Petch effect. Moreover, the quantities of both models are closely related. In particular, in both models the onset of plastic grain boundary yielding is controlled by grain boundary stresses. In the Eshelby model, these are shear stresses acting at dislocation sources

close to the grain boundary, while the gradient theory contains grain boundary microtractions.
In certain situations, the grain boundary stresses of both models are found to be related. However, they scale differently with the grain size. More precisely, the microtractions of the gradient model need to be rescaled by a factor of $1/\sqrt{D}$ to be a reasonable (but still rough) estimation of the Eshelby grain boundaries stresses.
Clearly, when applying a gradient theory, it is not desirable to explicitly account for a characteristic length of the system under consideration. Nevertheless, in order to verify the theoretical results, the aforementioned rescaling law is applied in three-dimensional periodic finite element simulations of microstructures consisting of several hundred grains. Indeed, the simulations match the Hall-Petch scaling law in a wide range of grain sizes.
Having obtained a numerically efficient and robust gradient plasticity model, the next part of the thesis is concerned with the replacement of the phenomenological work-hardening model. Conventional phenomenological work-hardening models usually neglect dislocation transport. Therefore, their application in micro-plasticity may yield unrealistic results when the free paths of the dislocations approach the order of magnitude of the system size, i.e., the transport is no longer negligible.
The starting point of the modification is the revision of Hochrainer's higher dimensional continuum dislocation theory. This kinematical framework provides partial differential evolution equations for the dislocation density as well as the average curvature. These have been derived from the averaging of systems of individual dislocations, i.e., they account for dislocation transport and curvature induced line length production.
After a detailed motivation and interpretation of the different evolution equation terms, a simplified version is discussed. Subsequently, the simplified equations are coupled to the equivalent plastic strain gradient plasticity theory that has been developed in the preceding part of the

thesis. Besides the aforementioned modification of the work-hardening model, the introduction of further generalized stresses, conjugate to the dislocation densities and the plastic slips, is discussed. The principal effect of this modification is a curvature related backstress in the flow rule as well as additional generalized force balance equations. A numerical example closes the chapter.
To summarize, several open questions in gradient plasticity theory have been worked on in this thesis. Clearly, several aspects of the work call for further investigations. Some examples are given in the following.

- The grain boundary yield criterion has a substantial influence on the overall model response. Yet, the grain boundary yield strength has been treated as material constant. However, in a more realistic setting, the influence of the grain misorientation on the grain boundary yield strength should be considered. Moreover, possible grain boundary jumps of the plastic field variable should be accounted for. This requires an extended constitutive model taking into account these jumps.
- Many crystal plasticity theories are based on the full dislocation density tensor. Compared to these, the equivalent plastic strain model implies a loss of information concerning the direction of the plastic flow. Therefore, clear guidelines are needed, indicating which situations allow for the usage of the equivalent plastic strain model and which applications require more elaborate frameworks. In the considered examples, the microstructures did not exhibit any preferred directions. The equivalent plastic strain model seems to yield reasonable results in this case. However, the applicability of the model to morphologies with distinct anisotropies remains to be quantified.
- Additional comparisons of the model response to experiments and DDD-simulations would be helpful to further investigate the accuracy of the model. In this context, plastically heterogeneous deformations are particularly interesting, since they facilitate a

further investigation of the rank-one and logarithmic energies, which have been applied in Chap. 2.

- The enhanced work-hardening model based on the dislocation density and the average curvature is able to account for dislocation transport. However, several questions arise in this context. For example, the kinematic framework does not yet account for the production of additional dislocations, i.e., the total number of dislocations is preserved. The same holds for dislocation annihilation.
- Physically meaningful dislocation flux terms need to be defined at open boundaries as well as grain boundaries. As long as this type of boundary and interface conditions is not available, a rigorous application of the continuum dislocation theory is limited to periodic problems with only one plastic phase. Therefore, the interactions of dislocations with grain boundaries need to be understood physically, before a meaningful mathematical theory can be developed.
- Substantial numerical problems arise, if the dislocation density approaches zero. In this case, the material behavior instantly changes from plastic to elastic which may lead to convergence problems. In addition, evaluating the second evolution equation of the dislocation continuum theory requires the division by the dislocation density. Note that the simplified version of the theory is not exact, i.e., even analytical solutions may contain negative dislocation density values. Therefore, robust numerical schemes are needed which prevent the dislocation density from becoming negative.
- Finally, the theory may be generalized to the geometrically nonlinear case.

Appendix A

A.1 Vanishing External Power of Micro Tractions

The second term of Eq. (2.36) reads

$$\int_{\Delta\mathcal{B}} \boldsymbol{m} \cdot \dot{\boldsymbol{H}}^{\mathrm{p}} \, \mathrm{d}a, \tag{A.1}$$

with $\dot{\boldsymbol{H}}^{\mathrm{p}} = \dot{\gamma} \boldsymbol{e}_1 \otimes \boldsymbol{e}_2$ and

$$\boldsymbol{m} \overset{(2.5)}{=} \boldsymbol{M}\hat{\boldsymbol{n}}. \tag{A.2}$$

The boundary is given by $\partial(\Delta\mathcal{B}) = A_1^+ \cup A_1^- \cup A_2^+ \cup A_2^- \cup A_3^+ \cup A_3^-$, where $A_i^\pm$ denote the surfaces of the cuboid $\Delta\mathcal{B}$ with associated surface normals $\boldsymbol{e}_i^\pm$ $(i = 1, 2, 3)$. At the top and bottom ($A_2^\pm$), the microtractions read

$$\begin{aligned} \boldsymbol{m}_{\mathrm{T}} &= \boldsymbol{M}(x_1)\hat{\boldsymbol{n}}_{\mathrm{T}}, \quad \hat{\boldsymbol{n}}_{\mathrm{T}} \overset{(2.6)}{=} -\epsilon \boldsymbol{e}_2 = -\hat{\boldsymbol{n}}_{\mathrm{B}} & \text{(A.3)} \\ \Rightarrow \boldsymbol{m}_{\mathrm{B}} &= \boldsymbol{M}(x_1)\hat{\boldsymbol{n}}_{\mathrm{B}} = -\boldsymbol{m}_{\mathrm{T}}. & \text{(A.4)} \end{aligned}$$

Since $\dot{\boldsymbol{H}}_{\mathrm{T}}^{\mathrm{p}} = \dot{\boldsymbol{H}}_{\mathrm{B}}^{\mathrm{p}}$, the top and bottom contributions ($A_2^\pm$) to the integral (A.1) cancel each other. The $A_3^\pm$-contributions vanish based on analogous arguments. The $A_1^\pm$-contributions vanish since $\dot{\boldsymbol{H}}^{\mathrm{p}} = \boldsymbol{0}$ in the elastic phase.

A.2 Power Balance for Rank-One Energy

With the representation $W = W_{\mathrm{e}} + W_{\mathrm{g}}$ (Eq. (2.10)), the power of internal forces (2.38) can be rewritten as

$$\mathcal{P}_{\mathrm{int}}(\Delta\mathcal{B}) = \int_{\Delta\mathcal{B}} \left(\partial_{\varepsilon} W_{\mathrm{e}} \cdot \dot{\varepsilon} + \partial_{\varepsilon^{\mathrm{p}}} W_{\mathrm{e}} \cdot \dot{\varepsilon}^{\mathrm{p}} + \dot{W}_{\mathrm{g}} + \mathcal{D} \right) \mathrm{d}v. \tag{A.5}$$

Since $W_\mathrm{e} = (\boldsymbol{\varepsilon} - \boldsymbol{\varepsilon}^\mathrm{p}) \cdot \mathbb{C}[\boldsymbol{\varepsilon} - \boldsymbol{\varepsilon}^\mathrm{p}]/2$, it follows that $\boldsymbol{\sigma} = \partial_{\boldsymbol{\varepsilon}} W_\mathrm{e} = -\partial_{\boldsymbol{\varepsilon}^p} W_\mathrm{e}$ and

$$\mathcal{P}_\mathrm{int}(\Delta\mathcal{B}) = \int_{\Delta\mathcal{B}} \left(\boldsymbol{\sigma}\cdot\dot{\boldsymbol{\varepsilon}} - \boldsymbol{\sigma}\cdot(\dot{\gamma}\,\mathrm{sym}(\boldsymbol{e}_1\otimes\boldsymbol{e}_2)) + \dot{W}_\mathrm{g} + (\tau_0^\mathrm{C} + \tau^\mathrm{D}\dot{\gamma}/\dot{\gamma}_0)\dot{\gamma}\right) \mathrm{d}v. \quad \text{(A.6)}$$

Here, Eq. (2.34) has been applied. From $\mathcal{P}_\mathrm{ext}(\Delta\mathcal{B}) = \mathcal{P}_\mathrm{int}(\Delta\mathcal{B})$ and Eq. (2.37), it follows that

$$\int_{-s/2-\epsilon}^{s/2+\epsilon} \left(-\sigma_{12}\,\dot{\gamma} + \dot{W}_\mathrm{g} + \left(\tau_0^\mathrm{C} + \tau^\mathrm{D}\frac{\dot{\gamma}}{\dot{\gamma}_0}\right)\dot{\gamma}\right) \mathrm{d}x_1 = 0. \quad \text{(A.7)}$$

The total defect energy rate is given by

$$\int_{-s/2-\epsilon}^{s/2+\epsilon} \dot{W}_\mathrm{g}\,\mathrm{d}x_1 \overset{(2.30)}{=} cGb\,\frac{\mathrm{d}}{\mathrm{d}t}\int_{-s/2-\epsilon}^{s/2+\epsilon} |\gamma'(x_1)|\,\mathrm{d}x_1 \quad \text{(A.8)}$$

$$\overset{(2.35)}{=} cGb\,\frac{\mathrm{d}}{\mathrm{d}t}\int_{-s/2-\epsilon}^{0} \gamma'(x_1)\,\mathrm{d}x_1 - cGb\,\frac{\mathrm{d}}{\mathrm{d}t}\int_{0}^{s/2+\epsilon} \gamma'(x_1)\,\mathrm{d}x_1 \quad \text{(A.9)}$$

$$= 2cGb\,\dot{\gamma}(0). \quad \text{(A.10)}$$

From this equation and Eq. (A.7), relation (2.39) immediately follows.

A.3 Sensitivity Analysis

According to Eq. (2.53) the macroscopic shear strain can be additively decomposed into an elastic part σ_{12}/G and a plastic part

$$\bar{\gamma}_\mathrm{p}/\alpha_0 = \frac{L_\tau}{e(s+h)}\left(\exp\left(\frac{s}{2L_\tau}\right)(s - 2L_\tau) + 2L_\tau\right), \quad \text{(A.11)}$$

with L_τ as defined in Eq. (2.52). The stress can be obtained as the inverse of (A.11), $\sigma_{12} = g(\bar{\gamma}_\mathrm{p}/\alpha_0)$. The function g is illustrated qualitatively in Fig. A.1.

The total differential reads

$$\mathrm{d}\sigma_{12} = \partial_{\bar{\gamma}_\mathrm{p}}\sigma_{12}\,\mathrm{d}\bar{\gamma}_\mathrm{p} + \partial_{\alpha_0}\sigma_{12}\,\mathrm{d}\alpha_0 = g'\,\mathrm{d}\left(\frac{\bar{\gamma}_\mathrm{p}}{\alpha_0}\right) \quad \text{(A.12)}$$

$$= \frac{1}{\alpha_0}g'\,\mathrm{d}\bar{\gamma}_\mathrm{p} - \frac{\bar{\gamma}_\mathrm{p}}{\alpha_0^2}g'\,\mathrm{d}\alpha_0. \quad \text{(A.13)}$$

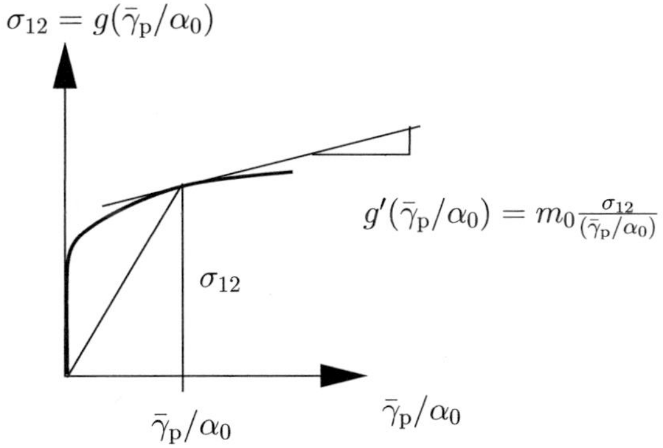

Figure A.1: Visualization of the function $g(\bar{\gamma}_{\mathrm{p}}/\alpha_0)$.

Hence, the sensitivity of σ_{12} with respect to changes of α_0 reads

$$\partial_{\alpha_0} \sigma_{12} = -\frac{\bar{\gamma}_{\mathrm{p}}}{\alpha_0^2} g'. \tag{A.14}$$

In the region of well-established plastic flow, one might conclude from Fig. A.1 that

$$\partial_{\alpha_0} \sigma_{12} = -\frac{\bar{\gamma}_{\mathrm{p}}}{\alpha_0^2} g' = -m_0 \frac{\bar{\gamma}_{\mathrm{p}}}{\alpha_0^2} \frac{\sigma_{12}}{(\bar{\gamma}_{\mathrm{p}}/\alpha_0)} = -m_0 \frac{\sigma_{12}}{\alpha_0}, \tag{A.15}$$

where m_0 takes a small value < 1. At constant plastic strain, small variations of α_0 lead to the following variations of σ_{12}

$$\Delta\sigma_{12} \approx -\sigma_{12}\, m_0 \frac{\Delta\alpha_0}{\alpha_0}. \tag{A.16}$$

Since m_0 is expected to be small, variations $|\Delta\alpha_0| < \alpha_0$ have a small influence on the overall size effect.

Appendix B

B.1 Displacement Consistency

Inserting the approximations (3.45) and (3.50) into Eq. (3.38) yields

$$\boldsymbol{u}(\boldsymbol{x}') = \frac{1}{\Delta V(\boldsymbol{x}')} \int\limits_{\Delta V(\boldsymbol{x}')} \Big(\underbrace{\boldsymbol{u}(\boldsymbol{x}) + \boldsymbol{H}(\boldsymbol{x})(\boldsymbol{x}'' - \boldsymbol{x}) - \Delta z(\boldsymbol{x}^{\mathrm{P}}(\boldsymbol{x}''))\gamma(\boldsymbol{x})\boldsymbol{d}}_{\boldsymbol{u}^*_{11}(\boldsymbol{x}'')} \Big) \, \mathrm{d}v''. \tag{B.1}$$

Neglecting boundary influences, this expression is given by

$$\boldsymbol{u}(\boldsymbol{x}') = \boldsymbol{u}(\boldsymbol{x}) + \boldsymbol{H}(\boldsymbol{x})(\boldsymbol{x}' - \boldsymbol{x}). \tag{B.2}$$

This means that the contribution of the fluctuations to the mesoscopic displacement approximation (inside of $V(\boldsymbol{x})$, Eq. (3.42)) indeed vanish. Therefore, the approach is consistent with the definition of $\boldsymbol{u}(\boldsymbol{x})$ (Eq. (3.38)).

B.2 Verification of W_{21}

Let Ω denote a volume (with centroid $\boldsymbol{x}_0$) with the same properties as ΔV. If Eq. (3.62) is correct, then the application of W_{21} is reasonable in situations where the micro-displacements within Ω can be approximated by

$$\boldsymbol{u}^*(\boldsymbol{x}) \approx \boldsymbol{u}(\boldsymbol{x}_0) + \boldsymbol{H}(\boldsymbol{x}_0)(\boldsymbol{x} - \boldsymbol{x}_0) + \frac{1}{2}(\nabla_{\mathrm{p}}\boldsymbol{H}(\boldsymbol{x}_0) : (\boldsymbol{x} - \boldsymbol{x}_0) \otimes (\boldsymbol{x} - \boldsymbol{x}_0)) - \Delta z(\boldsymbol{x})\gamma(\boldsymbol{x}_0)\boldsymbol{d}. \tag{B.3}$$

Here, the mesoscopic displacements $\boldsymbol{u}(\boldsymbol{x})$ are given by the first part. The last term represents the fluctuations $\tilde{\boldsymbol{u}}(\boldsymbol{x})$. The reasonability can be verified by noting that

$$\int\limits_{\Omega} \frac{1}{2} \boldsymbol{H}^*(\boldsymbol{x}) \cdot \mathbb{C}[\boldsymbol{H}^*(\boldsymbol{x})] \, \mathrm{d}v = \int\limits_{\Omega} W_{11}(\boldsymbol{x}) \, \mathrm{d}v. \tag{B.4}$$

B.3 Slip Consistency

The verification of Eq. (3.66) is carried out by proving the following identity

$$\boldsymbol{d} \cdot [\![\tilde{\boldsymbol{u}}_2(\boldsymbol{x}')]\!] = j_{\mathrm{u},2}(\boldsymbol{x}'), \quad \boldsymbol{x}' \in \mathcal{SB}. \tag{B.5}$$

Here, it has been assumed that the jump of the mesoscopic displacement vanishes, i.e. $[\![\boldsymbol{u}(\boldsymbol{x})]\!] = \boldsymbol{0}$.

$$\begin{aligned} & \boldsymbol{d} \cdot [\![\tilde{\boldsymbol{u}}_2(\boldsymbol{x}')]\!] && \text{(B.6)} \\ \overset{(3.64),(3.66)}{=} \; & -\boldsymbol{d} \cdot \left(j_{\mathrm{u},2}(\boldsymbol{x}' + h/2\,\boldsymbol{n})\boldsymbol{d}\frac{-h/2}{h} - j_{\mathrm{u},2}(\boldsymbol{x}' - h/2\,\boldsymbol{n})\boldsymbol{d}\frac{h/2}{h} \right) && \text{(B.7)} \\ = \; & j_{\mathrm{u}0} + \boldsymbol{b}_0 \cdot \big(\boldsymbol{x}' - \boldsymbol{x}\big) = j_{\mathrm{u},2}(\boldsymbol{x}'), \quad \boldsymbol{x}' \in \mathcal{SB}. && \text{(B.8)} \end{aligned}$$

B.4 Verification of W_{12}

The result (3.89) can be verified in the special case, where the micro displacement field $\boldsymbol{u}^*(\boldsymbol{x})$ can be represented based on the expression

$$\boldsymbol{u}^*(\boldsymbol{x}) = \boldsymbol{u}_0 + \boldsymbol{H}_0(\boldsymbol{x} - \boldsymbol{x}_0) - \big(j_{\mathrm{u}0} + \boldsymbol{b}_0 \cdot (\boldsymbol{x}^{\mathrm{p}}(\boldsymbol{x}) - \boldsymbol{x}_0)\big)\frac{\Delta z(\boldsymbol{x})}{h}\boldsymbol{d}. \tag{B.9}$$

This implies, that the mesoscopic displacement field $\boldsymbol{u}$ is given by a linear function in the averaging volume. As has been argued before (Eqns. (3.68), (3.70) and (3.72)),

$$\gamma(\boldsymbol{x}) = \gamma(\boldsymbol{x}_0) + \nabla_{\mathrm{p}}\gamma(\boldsymbol{x}_0) \cdot (\boldsymbol{x} - \boldsymbol{x}_0) = \frac{1}{h}\big(j_{\mathrm{u}0} + \boldsymbol{b}_0 \cdot (\boldsymbol{x} - \boldsymbol{x}_0)\big), \tag{B.10}$$

with $\nabla_{\mathrm{p}}\gamma(\boldsymbol{x}_0) = \boldsymbol{b}_0/h$ due to Eq. (3.65). Hence, the mesoscopic plastic slip is also approximated linearly in the glide plane and constant in $\boldsymbol{n}$-

direction. It can be shown that the total strain energy of a volume $V(\boldsymbol{x}_0)$ is then given by

$$
\begin{aligned}
W_{\text{tot}} &= \frac{1}{2} \int\limits_{V(\boldsymbol{x}_0)} \big(\boldsymbol{H}_0 - \gamma(\boldsymbol{x})\boldsymbol{d} \otimes \boldsymbol{n}\big) \cdot \mathbb{C}[\boldsymbol{H}_0 - \gamma(\boldsymbol{x})\boldsymbol{d} \otimes \boldsymbol{n}] \, \mathrm{d}v \\
&+ \frac{1}{2} \int\limits_{V(\boldsymbol{x}_0)} \frac{h^2}{12} \big(\boldsymbol{d} \otimes \nabla_{\mathrm{p}}\gamma(\boldsymbol{x}_0)\big) \cdot \mathbb{C}[\boldsymbol{d} \otimes \nabla_{\mathrm{p}}\gamma(\boldsymbol{x}_0)] \, \mathrm{d}v \\
&= \int\limits_{V(\boldsymbol{x}_0)} W_{12} \, \mathrm{d}v,
\end{aligned}
\tag{B.11}
$$

with $C^{\text{dn}} = (\boldsymbol{d} \otimes \boldsymbol{n}) \cdot \mathbb{C}[\boldsymbol{d} \otimes \boldsymbol{n}]$.

Appendix C

C.1 Energetical Interfaces

First, some geometrical assumptions are made. Therefore, a macroscopic cubic polycrystal is assumed to be loaded in tension. The geometrical properties of the microstructure are assumed to have no preferred direction and the grains are roughly approximated by spheres. The total volume of the polycrystal with edge length L reads

$$V = L^3 = \sum_i V_i \approx N_\circ V_\circ = \frac{\pi}{6} N_\circ D^3 = \frac{D}{3}\left(\frac{N_\circ}{2}\pi D^2\right), \tag{C.1}$$

where i is the grain index, $N_\circ \gg 1$ represents the total number of grains and D is the average grain diameter. $V_\circ$ is the volume of a sphere with diameter D. The total grain boundary surface is given by

$$\Gamma = \frac{1}{2}\sum_i \Gamma_i \approx \frac{N_\circ}{2} A_\circ = \frac{N_\circ}{2}\pi D^2 \tag{C.2}$$

$$\overset{(C.1)}{\Rightarrow} \Gamma \approx \frac{3V}{D}. \tag{C.3}$$

Here, Γ_i is the grain boundary area of an individual grain. The factor $1/2$ accounts for the fact that a grain boundary is shared by two grains. $A_\circ$ denotes the surface of a sphere with diameter D.
In a tensile test, two forces F (with opposite sign) are applied at two opposite faces of the cubic polycrystal. Assuming well-established plastic flow, the macroscopic strain rate reads

$$\dot{\bar{\varepsilon}} = \frac{v}{L}, \tag{C.4}$$

where v denotes the relative velocity between the two aforementioned faces.

In case of a very coarse-grained polycrystal (i.e. grain boundary influences are assumed negligible) the macroscopic tensile stress reads

$$\bar{\sigma} = \frac{F_{\text{coarse}}}{L^2}. \tag{C.5}$$

The power of the external forces is given by

$$P_0 = F_{\text{coarse}} v = \bar{\sigma}\dot{\bar{\varepsilon}} L^3. \tag{C.6}$$

This is the power of the external forces required to deform a coarse-grained polycrystal, where the grain boundary influence has been neglected.
In the following, also fine-grained polycrystals are considered. In order to phenomenologically model the resistance of the grain boundaries, the grain boundary energy density (6.2) is introduced

$$W_\Gamma = W_\Gamma(\varepsilon^{\text{p}}), \quad W_{\Gamma,\text{tot}} = \int_\Gamma W_\Gamma \, \text{d}a. \tag{C.7}$$

The introduction of W_Γ influences the macroscopic response of the polycrystal model. It is assumed that

$$F_{\text{fine}} v = (\bar{\sigma} + \Delta\bar{\sigma})\dot{\bar{\varepsilon}} L^3 \approx P_0 + \dot{W}_{\Gamma,\text{tot}}. \tag{C.8}$$

The stress increase $\Delta\bar{\sigma}$ is a consequence of the introduction of the grain boundary energy. It reads

$$\Delta\bar{\sigma} = \frac{\dot{W}_{\Gamma,\text{tot}}}{\dot{\bar{\varepsilon}} L^3}. \tag{C.9}$$

In the following, it is assumed that the macroscopic strain is sufficiently large such that the plastic flow is established in the whole grain ensemble. In this case, the average plastic strain rate can in most cases safely be expected to be nearly proportional to the macroscopic strain rate

$$\dot{\bar{\varepsilon}}^{p} = \frac{1}{V} \int_V \dot{\varepsilon}^{\text{p}} \, \text{d}v \approx k_\varepsilon \dot{\bar{\varepsilon}}, \tag{C.10}$$

where k_ε is a constant of proportionality. Furthermore, the rate of $W_{\Gamma,\text{tot}}$ is estimated by

$$\dot{W}_{\Gamma,\text{tot}} = \int_\Gamma W'_\Gamma(\varepsilon^{\text{p}}) \, \dot{\varepsilon}^{\text{p}} \, \text{d}a \approx W'_\Gamma(\bar{\varepsilon}^{\text{p}}) \, \dot{\bar{\varepsilon}}^{p} \, \Gamma. \tag{C.11}$$

Insertion of Eqns. (C.10) and (C.11) into Eq. (C.9) yields

$$\Delta\bar{\sigma} = \frac{k_\varepsilon W'_\Gamma(\bar{\varepsilon}^{\mathrm{p}})\,\dot{\bar{\varepsilon}}\,\Gamma}{\dot{\bar{\varepsilon}} L^3} \overset{(C.3)}{=} \frac{3k_\varepsilon W'_\Gamma(\bar{\varepsilon}^{\mathrm{p}})}{D}. \tag{C.12}$$

Based on the assumptions made, it is found that the increase in yield stress is proportional to the inverse of the grain diameter

$$\Delta\bar{\sigma} \sim \frac{1}{D}. \tag{C.13}$$

C.2 Dissipative Interfaces

The analogue of Eq. (C.8) reads

$$L^3(\bar{\sigma} + \Delta\bar{\sigma})\dot{\bar{\varepsilon}} = P_0 + D_{\Gamma,\mathrm{tot}} \quad \Rightarrow \Delta\bar{\sigma} = \frac{D_{\Gamma,\mathrm{tot}}}{\dot{\bar{\varepsilon}} L^3}. \tag{C.14}$$

Here, $D_{\Gamma,\mathrm{tot}}$ denotes the total dissipation at the grain boundaries. $D_{\Gamma,\mathrm{tot}}$ is assumed to be proportional to the total grain boundary area Γ on the one hand and the plastic strain rate $\dot{\bar{\varepsilon}}^p$ on the other hand

$$D_{\Gamma,\mathrm{tot}} = \int_\Gamma \mathcal{D}_\Gamma \,\mathrm{d}a = \int_\Gamma \Xi^{\mathrm{d}}_{\mathrm{GB}}\,\dot{\varepsilon}^{\mathrm{p}}\,\mathrm{d}a \approx \int_\Gamma \Xi^{\mathrm{d}}_\Gamma \,\mathrm{d}a\,\dot{\bar{\varepsilon}}^p \approx k_D \Gamma\,\dot{\bar{\varepsilon}}^p, \tag{C.15}$$

where k_D is a constant of proportionality, $\mathcal{D}_\Gamma$ denotes the dissipation per unit (grain boundary) area and Ξ^{d}_Γ is a microstress which extends power over $\dot{\varepsilon}^{\mathrm{p}}$. Since the dissipative grain boundary does not store energy, this power cannot be recovered and Ξ^{d}_Γ is purely dissipative. The application of Eqns. (C.3) and (C.10) yields

$$D_{\Gamma,\mathrm{tot}} = k_D\,\Gamma\,\dot{\bar{\varepsilon}}^p \overset{(C.10)}{=} k_D\,\Gamma\,k_\varepsilon \dot{\bar{\varepsilon}} \overset{(C.3)}{=} \frac{3L^3\,k_D\,k_\varepsilon \dot{\bar{\varepsilon}}}{D}. \tag{C.16}$$

A comparison with Eq. (C.14) yields

$$\Delta\bar{\sigma} = \frac{3k_\varepsilon k_D}{D}. \tag{C.17}$$

If k_ε is assumed independent of the grain diameter D, the result does, again, not match the Hall-Petch relation (6.1).

C.3 Interpretation of the Material Parameters

In the following, the influence of various material parameters on the overall behavior are discussed. First, a material with large grains is considered. In this case, the grain boundary influence is negligible. At room temperature the rate sensitivity usually takes high values (e.g. $p = 20$). Then, the simulation is rather insensitive with respect to changes of the reference shear rate $\dot{\gamma}_0$ (at least at moderate up to high strain rates). Consequently, the overall yield stress is expected to be approximately given by

$$\sigma_{Y0} \approx M_{\mathrm{T}}(\tau_0^{\mathrm{C}} + \tau^{\mathrm{D}}), \tag{C.18}$$

where M_{T} is a Taylor factor ($M_{\mathrm{T}} \approx 3.06$ for FCC metals, Taylor (1938)). Cold work hardening is accounted for by the isotropic hardening stress $\beta = \partial_{\gamma_{\mathrm{eq}}} W_{\mathrm{h}}$ (Eq. (5.18)).

For small grains, the influence of the grain boundaries must be taken into account. The grain boundary behavior is governed by the material constant $k^{\mathrm{C}} = \tau_{\mathrm{s}}^{\mathrm{C}}/(4k_1)$, see Eqns. (6.21) and (6.22). In the following, it is shown that the Hall-Petch slope k_{HP}, Eq. (6.1), is expected to be directly proportional to k^{C}. In appendix C.2, the following estimation is derived

$$\Delta\bar{\sigma} = \frac{3k_\varepsilon k_D}{D}. \tag{C.19}$$

with $k_\varepsilon = \dot{\bar{\gamma}}_{\mathrm{eq}}/\dot{\bar{\varepsilon}}$ (volume averages) and

$$k_D \overset{(\mathrm{C.15})}{=} \frac{1}{\Gamma}\int_\Gamma \Xi_\Gamma \, da \approx k^{\mathrm{C}}\sqrt{D}. \tag{C.20}$$

Here, Eqns. (6.21), (6.22) and $f_\Gamma^* = 0$ have been exploited. Hence, this relation is assumed to hold in the case of well established plastic flow. It follows that the Hall-Petch slope is directly proportional to k^{C}

$$\Delta\bar{\sigma} = \frac{3k_\varepsilon k^{\mathrm{C}}}{\sqrt{D}} = \frac{k_{\mathrm{HP}}}{\sqrt{D}}. \tag{C.21}$$

This result allows for a first guess of k^{C} without having to fit any numerical model

$$k_{\mathrm{HP}} \approx 3k_\varepsilon k^{\mathrm{C}} \Leftrightarrow k^{\mathrm{C}} \approx \frac{k_{\mathrm{HP}}}{3k_\varepsilon}. \tag{C.22}$$

This guess does not yet take into account the dependence of k^{C} on the grain misorientation. The parameter k_ε has been estimated from a numerical simulation. It was found that $k_\varepsilon \approx 3$ (for a set of parameters, which is typical for copper).
Once, the simulations have been done for a given value of k^{C}, the Hall-Petch slope k_{HP} can be manipulated by changing k^{C}. Multiplication of k^{C} by a given factor will roughly in- or decrease the Hall-Petch slope k_{HP} by this factor (Eq. (C.22)).
Besides the parameter k^{C}, the influence of the defect energy $W_{\mathrm{g}}(\nabla\gamma_{\mathrm{eq}})$ remains to be discussed. For the simulations, a quadratic approach was applied

$$W_{\mathrm{g}}(\nabla\gamma_{\mathrm{eq}}) = \frac{1}{2} K_{\mathrm{G}} \nabla\gamma_{\mathrm{eq}} \cdot \nabla\gamma_{\mathrm{eq}}. \tag{C.23}$$

Interestingly, in Sect. 5.3.3 (Fig. 5.4) it has been found that the variation of the parameter K_{G} (in a quite wide range) had almost no effect on the macroscopic stress strain curves. However, if K_{G} is chosen sufficiently small, the elasto-plastic transition of large-grain-systems becomes smoother. In this work, the value $K_{\mathrm{G}} = 0.01$ N is chosen.
Many authors introduce an internal length scale instead of a parameter like K_{G} by setting $K_{\mathrm{G}} = l^2 E$, where E is Young's modulus of the macroscopic material. Taking copper with $E \approx 110$ GPa, the internal length scale is given by $l = \sqrt{K_{\mathrm{G}}/E} \approx 0.3\mu\mathrm{m}$.

Appendix D

Proof of Eq. 7.6

For simplicity, the index i of the dislocation is dropped in the proof. The derivatives of the direction vectors along the dislocation are given by

$$\frac{\partial \boldsymbol{e}_{\mathrm{l}}}{\partial s} = -\frac{1}{R}\boldsymbol{e}_{\nu}, \quad \frac{\partial \boldsymbol{e}_{\nu}}{\partial s} = \frac{1}{R}\boldsymbol{e}_{\mathrm{l}}, \tag{D.1}$$

where the local radius of curvature R is signed.

An intermediate configuration for one fixed moment $\tilde{t} \in [0; t]$ is introduced

$$\tilde{s} = s(S, \tilde{t}), \quad \boldsymbol{x}_{\tilde{t}} = \boldsymbol{x}(S(\tilde{s}, \tilde{t}), t) = \boldsymbol{x}_{\tilde{t}}(\tilde{s}, t). \tag{D.2}$$

At the point P, a generalized tangent vector is defined

$$\boldsymbol{e}_{\mathrm{l}}^{\tilde{t}} = \frac{\partial \boldsymbol{x}_{\tilde{t}}}{\partial \tilde{s}} \quad \Rightarrow \boldsymbol{e}_{\mathrm{l}} = \boldsymbol{e}_{\mathrm{l}}^{t}. \tag{D.3}$$

Now the case $\tilde{t} = t$ is considered, i.e., $\dot{\tilde{s}} = v_{\mathrm{T}}$. The time rate of change of the generalized tangent vector is evaluated as follows

$$\left.\frac{\mathrm{d}\boldsymbol{e}_{\mathrm{l}}^{\tilde{t}}}{\mathrm{d}t}\right|_{\tilde{t}=t} = \left.\frac{\partial \boldsymbol{e}_{\mathrm{l}}^{\tilde{t}}}{\partial \tilde{s}}\dot{\tilde{s}}\right|_{\tilde{t}=t} + \left.\frac{\partial \boldsymbol{e}_{\mathrm{l}}^{\tilde{t}}}{\partial t}\right|_{\tilde{t}=t} = -\frac{1}{R}\boldsymbol{e}_{\nu} v_{\mathrm{T}} + \left.\frac{\partial \boldsymbol{e}_{\mathrm{l}}^{\tilde{t}}}{\partial t}\right|_{\tilde{t}=t}. \tag{D.4}$$

The first term results from the motion of P in the direction $\boldsymbol{e}_{\mathrm{l}}$, the second term from the motion of $\mathcal{C}$.

The second term, evaluated at $\tilde{t} = t$, and rearranged using the chain rule yields

$$\left.\frac{\partial \boldsymbol{e}_{\mathrm{l}}^{\tilde{t}}}{\partial t}\right|_{\tilde{t}=t} = \left.\frac{\partial}{\partial \tilde{s}}\frac{\partial \boldsymbol{x}_{\tilde{t}}}{\partial t}\right|_{\tilde{t}=t} = \frac{\partial \boldsymbol{\nu}}{\partial s} = \frac{\partial \nu}{\partial s}\boldsymbol{e}_{\nu} + \frac{\nu}{R}\boldsymbol{e}_{\mathrm{l}}. \tag{D.5}$$

If the point P is enforced to indicate a constant tangent, its velocity can be deduced from the requirement

$$\left.\frac{\mathrm{d}\boldsymbol{e}_{\mathrm{I}}^{\tilde{t}}}{\mathrm{d}t}\right|_{\tilde{t}=t} \cdot \boldsymbol{e}_{\nu} \stackrel{!}{=} 0, \tag{D.6}$$

which leads to the following expression for the tangential velocity of P

$$v_{\mathrm{T}} = -R\dot{\vartheta}. \tag{D.7}$$

List of Frequently Used Symbols

Greek letters

α_0	characteristic density of statistically stored dislocations
α_{L}	transition density
$\boldsymbol{\alpha}^{\mathrm{Tr}}$	trial dislocation density tensor
$\boldsymbol{\alpha}$	dislocation density tensor
β	energetic isotropic hardening stress
$\delta_{\alpha\beta}$	Kronecker delta
$\delta_{\mathcal{SB}}$	slip band delta function
$\dot{\varepsilon}_0$	reference strain rate
ϵ	small number
η	generalized stress conjugate to $\dot{\rho}$
$\boldsymbol{\varepsilon}^{\mathrm{e}}$	elastic strain tensor
$\boldsymbol{\varepsilon}^{\mathrm{p}}$	plastic strain tensor
$\boldsymbol{\varepsilon}_i$	strain of layer with index i
$\boldsymbol{\epsilon}$	permutation (Ricci's) tensor
$\boldsymbol{\varepsilon}$	infinitesimal strain tensor
$\bar{\gamma}$	macroscopic shear strain
$\dot{\gamma}_0$	reference slip rate
$\dot{\gamma}_{\mathrm{L}}$	regularization parameter
γ_0	characteristic plastic shear parameter
γ_{eq}	equivalent plastic strain
γ	plastic shear strain

$\boldsymbol{\kappa}$	dislocation density vector
κ	norm of $\boldsymbol{\kappa}$
$\dot{\lambda}_0$	reference strain rate
Λ	overstress type function
λ	plastic slip parameter
$\mathring{\lambda}$	Lagrange multiplier
μ	history variable
ϕ_v	dissipation potential w.r.t. the relative layer velocity
$\Pi_{\mathcal{D}}$	integral dissipation potential
Π_L	integral Lagrange parameter potential
Π_W	integral stored energy potential
Π_{ext}	potential of external forces
Π	integral potential
φ_{g}^{1}	indicator function of rank one energy
$\varphi_{\mathrm{g}}^{1,\mathrm{Tr}}$	trial indicator function
φ	dislocation orientation
$\rho_{\odot}$	geometrically necessary screw dislocation density
ρ_{φ}	orientation-dependent dislocation density
$\rho_{\vdash}$	geometrically necessary edge dislocation density
ρ_m	mass density
ρ_{GND}	geometrically necessary dislocation density
ρ_{sat}	saturation dislocation density
ρ_{SSD}	statistically stored dislocation density
ρ	total dislocation density
$\boldsymbol{\sigma}$	Cauchy stress tensor
σ_0^{C}	initial yield stress
σ^{D}	drag stress
σ_{b}	bending stress
σ_{Y0}	yield stress of a coarse grained polycrystal
σ_{Y}	yield stress

σ_{ij}	components of the Cauchy stress w.r.t. a Cartesian basis
τ^{B}	backstress
τ_0^{C}	initial critical resolved shear stress (initial yield stress)
τ_∞^{C}	saturation yield stress
τ^{C}	critical resolved shear stress
τ^{D}	drag stress
τ^{d}	dissipative shear stress
τ^{eff}	effective shear stress
τ^{k}	generalized stress conjugate to $\dot{\gamma}$
τ_{ext}	externally applied shear stress
τ_{f}	lattice friction stress
$\tau_{\mathrm{s}}^{\mathrm{C}}$	critical dislocation source stress
τ_{s}	dislocation source stress
τ_{t}	torsion shear stress
τ	resolved shear stress
Θ_0	initial hardening modulus
θ_0	deformation angle
ϑ	angular dislocation velocity
υ	Poisson's ratio
$\boldsymbol{\nu}$	dislocation velocity vector
ν_0	dislocation reference velocity
ν	dislocation velocity
ω	defect energy exponent
$\boldsymbol{\xi}$	microstress
Ξ_0^{C}	initial grain boundary yield strength
Ξ_Γ	grain boundary microtraction
ξ	scalar microstress
ζ	phenomenological plastic field variable

Latin letters

$\hat{A}$ jacobian matrix
a_{T} constant of order one
A set of surface points/area
a length of bar
$\boldsymbol{b}_0$ gradient of linear approximation of j_{u}
$\boldsymbol{b}_{\mathrm{tot}}$ total or net Burgers vector
$\boldsymbol{b}$ Burgers vector
b absolute value of Burgers vector
$\mathbb{C}$ elastic stiffness tensor of fourth order
$\tilde{c}$ yield resistance
C^{ij} projection of the stiffness tensor ($i, j \in \{\mathrm{d}, \mathrm{h}, \mathrm{n}\}$)
c_0 log. energy constant
c_1 constant of order unity
c_ω defect energy material parameter
c_ρ initial density modulus
C_{ijkl} stiffness tensor component
c constant of order unity
$\mathcal{D}$ volumetric dissipation density
$\boldsymbol{d}$ slip direction
$\hat{D}$ algorithmic tangent matrix
d_s dislocation stretch rate
D_{g} edge length of square grains
D_{tot} total (integral) dissipation
D mean grain diameter
d diameter of specific grain
$\boldsymbol{e}_\nu$ dislocation velocity direction
$\boldsymbol{e}_i$ orthonormal unit vectors ($i = 1, 2, 3$)
$\boldsymbol{e}_x$ unit vector in x-direction
$\boldsymbol{e}_y$ unit vector in y-direction

$\boldsymbol{e}_z$	unit vector in z-direction
$\boldsymbol{e}_\mathrm{l}$	line direction
E_S	quasi-material dislocation strain measure
E	Young's modulus
e	Euler's constant
$\mathcal{F}_\mathrm{L}$	Lagrangian
$\boldsymbol{f}_m$	discrete mass flux
$\tilde{f}$	equivalent stress
f_Γ^*	grain boundary yield criterion in terms of τ_s
f_Γ	grain boundary yield criterion
F_S	quasi-material dislocation stretch measure
$f_{\alpha\beta}$	history variable function
f_c	compressive yield criterion
f_t	tensile yield criterion
F	concentrated force
f	yield criterion
G	shear modulus
$\boldsymbol{H}^\mathrm{e}$	elastic part of the displacement gradient
$\boldsymbol{H}^\mathrm{p}$	plastic distorsion
$\boldsymbol{H}$	displacement gradient
$\boldsymbol{h}$	crystallographic vector, cross product of $\boldsymbol{d}$ and $\boldsymbol{n}$
H_χ^M	penalty parameter
H_χ^s	penalty parameter
H_χ	micromorphic coupling modulus
$h_{\alpha\beta}$	hardening moduli
H	hardening modulus
h	width of hard layer
I_T	torsion moment of inertia
j_u0	constant approximation of j_u
j_u	smooth displacement jump interpolation function

k_1	constant of proportionality
K_{G}	gradient modulus
k_{HP}	Hall-Petch slope
k_{L}	regularization parameter
k	dislocation curvature
Δl_{φ}	total segment line length associated to orientation φ
$\boldsymbol{l}$	crystallographic vector, cross product of $\boldsymbol{n}$ and $\boldsymbol{d}$
$l_{\mathcal{L}}$	segment length
L_{τ}	stress dependent length parameter
L_{c}	cube width
l_{s}	internal length scale parameter
L	characteristic length of the average volume
l	internal length scale parameter
Δm	total particle mass
$\boldsymbol{M}^{\mathrm{old}}$	(unconverged) approximation of $\boldsymbol{M}$
$\boldsymbol{M}^{\mathrm{Tr}}$	generalized trial stress
$\boldsymbol{M}$	generalized stress conjugate to $\dot{\boldsymbol{\alpha}}$
$\boldsymbol{m}$	microtraction
$\boldsymbol{P}^{\mathrm{S}}$	symmetric part of Schmid tensor
M_{φ}	number of discrete orientations
M_{b}	bending moment
m_{b}	distributed bending moment
M_{T}	Taylor factor
m_{t}	distributed torsion moment
M	non-vanishing component of $\boldsymbol{M}$
m	mass
$\boldsymbol{N}^{\mathrm{Tr}}$	trial direction
$\boldsymbol{n}$	normal vector
$\hat{\boldsymbol{n}}$	axial tensor
$\hat{N}$	large number of particles

n	preceding time step
$\tilde{\boldsymbol{n}}$	surface normal
N_φ	number of tangent points
$N_{\mathcal{L}}$	number of layers
$N_{\mathcal{SB}}$	number of slip bands
N_{c}	number of layers of cube
N_{g}	number of layers of grain
N	number of slip systems
n	orientation-dependent dislocation segment density
$\breve{p}$	micromorphic coupling stress
$\mathcal{P}_{\mathrm{ext}}$	power of external forces
$\mathcal{P}_{\mathrm{int}}$	power of internal forces
p_{int}	power density of internal forces
p	rate sensitivity parameter
q_α	isotropic hardening stress
q_{l}	latent hardening parameter
q	quadratic strain gradient function
ΔR	total radius of curvature
$\boldsymbol{r}^\sigma$	residual associated to Cauchy stress
r^p	residual associated to coupling stress $\breve{p}$
R_φ	radius of curvature
$\mathbb{S}$	compliance tensor
$\boldsymbol{s}^{\mathrm{old}}$	(unconverged) approximation of $\boldsymbol{s}$
$\boldsymbol{s}$	generalized stress conjugate to $\dot{\boldsymbol{H}}^{\mathrm{p}}$
s	width of soft phase
$\boldsymbol{t}$	traction vector
t	time
$\boldsymbol{u}_0$	constant term of Taylor series expansion of $\boldsymbol{u}^*$
$\boldsymbol{u}$	displacement field
u	displacement in x-direction

$\boldsymbol{v}_P$ velocity of point P
$\boldsymbol{v}_\mathrm{T}$ tangential velocity
$\boldsymbol{v}$ velocity
v_0 reference velocity
v_rel relative layer velocity
v displacement in y-direction
$\overline{W}_\mathrm{g}$ average volumetric energy
W_0 offset energy
W_χ micromorphic coupling energy
W_Γ grain boundary energy density (hardening potential)
w_ρ test function
W_i elastic energy of layer with index i
$W_{\Gamma,\mathrm{tot}}$ total (integral) grain boundary energy
$w_{\rho k}$ test function
W_e elastic strain energy density
W_g gradient hardening potential
W_h isotropic hardening potential
W_{tot} total elastic energy
$\boldsymbol{x}^\mathrm{P}$ orthogonal layer center plane projection
Δz signed distance from layer center plane

Sets

$\mathcal{A}$ active set of grain boundary nodes
$\mathcal{B}_i$ grain with index i
$\mathcal{B}$ body (set of material points)
$\mathcal{C}$ dislocation line
Γ_act plastically active part of grain boundaries
Γ set of all grain boundaries
$\mathcal{L}_i$ center plane of layer i
$\mathcal{L}_{\Delta\varphi}$ non-quasi-material dislocation segment

$\mathcal{L}$ dislocation segment
$\Delta\Omega$ averaging volume
P point with restricted tangent
$\mathbb{R}$ set of real numbers
$\mathcal{SB}_i$ slip band i
$\mathcal{SB}$ union of all slip bands
ΔV averaging volume

Special Operators

$\delta(\bullet)$ variation/virtual quantity
$\dot{(\bullet)}$ time rate of change/partial derivative w.r.t. time
$[\![\bullet]\!]$ jump of field quantity
$\nabla_{\mathrm{p}}(\bullet)$ projected gradient
$\otimes$ dyadic product

Sub- and superscripts

$(\bullet)^*$ microscopic quantity
$(\bullet)^{\mathrm{d}}$ dissipativie part
$(\bullet)^{\mathrm{e}}$ energetic/elastic part
$(\bullet)^{\mathrm{p}}$ plastic part
$(\bullet)_\alpha$ quantity associated to slip system α
$\bar{(\bullet)}$ given traction-quantity at Neumann boundary
$\overline{(\bullet)}$ average quantity
$\tilde{(\bullet)}$ fluctuation

Bibliography

Acharya, A., Bassani, J. L., 2000. Lattice incompatibility and a gradient theory of crystal plasticity. Journal of the Mechanics and Physics of Solids 48, 1565–1595.

Aifantis, E. C., 1987. The physics of plastic deformation. International Journal of Plasticity 3 (3), 211–247.

Aifantis, K., Senger, J., Weygand, D., Zaiser, M., 2009. Discrete dislocation dynamics simulation and continuum modeling of plastic boundary layers in tricrystal micropillars. In: IOP Conference Series: Materials Science and Engineering. Vol. 3. IOP Publishing, p. 012025.

Aifantis, K., Soer, W., De Hosson, J., Willis, J., 2006. Interfaces within strain gradient plasticity: Theory and experiments. Acta materialia 54 (19), 5077–5085.

Aifantis, K., Willis, J., 2005. The role of interfaces in enhancing the yield strength of composites and polycrystals. Journal of the Mechanics and Physics of Solids 53 (5), 1047–1070.

Aifantis, K. E., Weygand, D., Motz, C., Nikitas, N., Zaiser, M., 2012. Modeling microbending of thin films through discrete dislocation dynamics, continuum dislocation theory, and gradient plasticity. Journal of Materials Research 27 (03), 612–618.

Anand, L., Aslan, O., Chester, S. A., 2012. A large-deformation gradient theory for elastic–plastic materials: strain softening and regularization of shear bands. International Journal of Plasticity 30, 116–143.

Anand, L., Gurtin, M., Lele, S., Gething, C., 2005. A one-dimensional theory of strain-gradient plasticity: formulation, analysis, numerical results. Journal of the Mechanics and Physics of Solids 53 (8), 1789–1826.

Armstrong, R. W., Codd, I., Douthwaite, R. M., Petch, N. J., 1962. The plastic deformation of polycrystalline aggregates. Philosophical Magazine 7 (73), 45–58.

Arsenlis, A., Parks, D. M., 2002. Modeling the evolution of crystallographic dislocation density in crystal plasticity. Journal of the Mechanics and Physics of Solids 50 (9), 1979–2009.

Asaro, R., 1975. Elastic-plastic memory and kinematic-type hardening. Acta metallurgica 23 (10), 1255–1265.

Ashby, M., 1970. The deformation of plastically non-homogeneous materials. Philosophical Magazine 21 (170), 399–424.

Aslan, O., Cordero, N. M., Gaubert, A., Forest, S., 2011. Micromorphic approach to single crystal plasticity and damage. International Journal of Engineering Science 49, 1311–1325.

Aslan, O., Forest, S., 2011. The micromorphic versus phase field approach to gradient plasticity and damage with application to cracking in metal single crystals. In: Multiscale Methods in Computational Mechanics. Springer, Dordrecht Heidelberg London New York, pp. 135–153.

Bardella, L., Segurado, J., Panteghini, A., Llorca, J., 2013. Latent hardening size effect in small-scale plasticity. Modelling and Simulation in Materials Science and Engineering 21 (5), 055009.

Bargel, H.-J., Schulze, G., 2008. Werkstoffkunde. Springer, Berlin Heidelberg.

Bargmann, S., Ekh, M., Runesson, K., Svendsen, B., 2010. Modeling of polycrystals with gradient crystal plasticity: A comparison of strategies. Philosophical Magazine 90, 1263–1288.

Baskaran, R., Akarapu, S., Mesarovic, S. D., Zbib, H. M., 2010. Energies and distributions of dislocations in stacked pile-ups. International Journal of Solids and Structures 47 (9), 1144–1153.

Bayley, C., Brekelmans, W., Geers, M., 2006. A comparison of dislocation induced back stress formulations in strain gradient crystal plasticity. International Journal of Solids and Structures 43, 7268–7286.

Becker, M., Jul. 2006. Incompatibility and instability based size effects in crystals and composites at finite elastoplastic strains. Ph.D. thesis, Universität Stuttgart.

Berdichevsky, V., 2006a. On thermodynamics of crystal plasticity. Scripta Mat. 54, 711–716.

Berdichevsky, V., Le, K., 2007. Dislocation nucleation and work hardening in anti-plane constrained shear. Continuum Mechanics and Thermodynamics 18 (7-8), 455–467.

Berdichevsky, V. L., 2006b. Continuum theory of dislocations revisited. Continuum Mechanics and Thermodynamics 18 (3-4), 195–222.

Bilby, B. A., Bullough, R., Smith, E., 1955. Continuous distributions of dislocations: a new application of the methods of non-Riemannian geometry. Proceedings of the Royal Society London A 231, 263–273.

Bittencourt, E., Needleman, A., Gurtin, M. E., der Giessen, E. V., 2003. A comparison of nonlocal continuum and discrete dislocation plasticity predictions. Journal of the Mechanics and Physics of Solids 51, 281–310.

Bortoloni, P., Cermelli, L., 2000. Statistically stored dislocations in rate-independent plasticity. In: Geometry, Continua and Microstuctures, I. Vol. 58. Universitae Politecnico di Torino, pp. 25–36.

Cermelli, P., Gurtin, M. E., 2002. Geometrically necessary dislocations in viscoplastic single crystals and bicrystals undergoing small deformations. International Journal of Solids and Structures 39, 6281–6309.

Chang, Y. W., Asaro, R. J., 1981. An experimental study of shear localization in aluminum-copper single crystals. Acta Metallurgica 29 (1), 241–257.

Cheong, K., Busso, E., Arsenlis, A., 2005. A study of microstructural length scale effects on the behaviour of FCC polycrystals using strain gradient concepts. International Journal of Plasticity 21 (9), 1797–1814.

Cleveringa, H., Van der Giessen, E., Needleman, A., 1997. Comparison of discrete dislocation and continuum plasticity predictions for a composite material. Acta Materialia 45 (8), 3163–3179.

Cordero, N., Forest, S., Busso, E. P., 2013. Micromorphic modelling of grain size effects in metal polycrystals. GAMM-Mitteilungen 36 (2), 186–202.

Cordero, N. M., Forest, S., Busso, E., Berbenni, S., Cherkaoui, M., 2012. Grain size effects on plastic strain and dislocation density tensor fields in metal polycrystals. Computational Materials Science 52, 7–13.

Cordero, N. M., Gaubert, A., Forest, S., Busso, E., Gallerneau, F., Kruch, S., 2010. Size effects in generalised continuum crystal plasticity for two-phase laminates. Journal of the Mechanics and Physics of Solids 58, 1963–1994.

Cuitiño, A. M., Ortiz, M., 1992. Computational modelling of single crystals. Modelling and Simulation in Materials Science and Engineering 1, 225–263.

Dahlberg, C. F., Faleskog, J., Niordson, C. F., Legarth, B. N., 2013. A deformation mechanism map for polycrystals modeled using strain gradient plasticity and interfaces that slide and separate. International Journal of Plasticity 43, 177–195.

de Borst, R., Mühlhaus, H.-B., 1992. Gradient-dependent plasticity: formulation and algorithmic aspects. Int. J. Numer. Meth. Engng. 35, 521–539.

Dimiduk, D., Uchic, M., Parthasarathy, T., 2005. Size-affected single-slip behavior of pure nickel microcrystals. Acta Materialia 53 (15), 4065–4077.

Djoko, J., Ebobisse, F., McBride, A., Reddy, B., 2007a. A discontinuous galerkin formulation for classical and gradient plasticity–part 1: Formulation and analysis. Computer Methods in Applied Mechanics and Engineering 196 (37), 3881–3897.

Djoko, J., Ebobisse, F., McBride, A., Reddy, B., 2007b. A discontinuous galerkin formulation for classical and gradient plasticity. Part 2: Algorithms and numerical analysis. Computer Methods in Applied Mechanics and Engineering 197 (1), 1–21.

Dunstan, D., Bushby, A., 2014. Grain size dependence of the strength of metals: The hall–petch effect does not scale as the inverse square root of grain size. International Journal of Plasticity 53, 56–65.

Engelen, R. A., Geers, M. G., Baaijens, F., 2003. Nonlocal implicit gradient-enhanced elasto-plasticity for the modelling of softening behaviour. International Journal of Plasticity 19 (4), 403–433.

Eringen, A. C., Suhubi, E., 1964. Nonlinear theory of simple micro-elastic solids–I. International Journal of Engineering Science 2 (2), 189–203.

Ertürk, I., van Dommelen, J., Geers, M., 2009. Energetic dislocation interactions and thermodynamical aspects of strain gradient crystal plasticity theories. Journal of the Mechanics and Physics of Solids 57, 1801–1814.

Eshelby, J., Frank, F., Nabarro, F., 1951. XLI. the equilibrium of linear arrays of dislocations. Philosophical Magazine 42 (327), 351–364.

Essmann, U., Rapp, M., 1973. Slip in copper crystals following weak neutron bombardment. Acta Metallurgica 21, 1305–1317.

Essmann, U., Mughrabi, H., 1979. Annihilation of dislocations during tensile and cyclic deformation and limits of dislocation densities. Philosophical Magazine 40 (6), 731–756.

Estrin, Y., 1996. Dislocation-density related constitutive modelling. In: Unified Constitutive Laws of Plastic Deformation. Academic Press, San Diego, pp. 69–106.

Evers, L. P., Brekelmans, W. A. M., Geers, M. G. D., 2004a. Non-local crystal plasticity model with intrinsic SSD and GND effects. Journal of the Mechanics and Physics of Solids 52, 2379–2401.

Evers, L. P., Brekelmans, W. A. M., Geers, M. G. D., 2004b. Scale dependent crystal plasticity framework with dislocation density and grain boundary effects. International Journal of Solids and Structures 41, 5209–5230.

Evers, L. P., Parks, D. M., Brekelmans, W. A. M., Geers, M. G. D., 2002. Crystal plasticity model with enhanced hardening by geometrically necessary dislocation accumulation. Journal of the Mechanics and Physics of Solids 50, 2403–2424.

Fleck, N. A., Hutchinson, J. W., 1993. A phenomenological theory for strain gradient effects in plasticity. Journal of the Mechanics and Physics of Solids 41, 1825–1857.

Fleck, N. A., Muller, G. M., Ashby, M. F., Hutchinson, J. W., 1994. Strain gradient plasticity: Theory and experiment. Acta metall. mater. 42 (2), 475–487.

Fleck, N. A., Willis, J. R., 2009a. A mathematical basis for strain-gradient plasticity theory–Part I: Scalar plastic multiplier. Journal of the Mechanics and Physics of Solids 57, 161–177.

Fleck, N. A., Willis, J. R., 2009b. A mathematical basis for strain-gradient plasticity theory. Part II: Tensorial plastic multiplier. Journal of the Mechanics and Physics of Solids 57 (7), 1045–1057.

Forest, S., 2009. Micromorphic approach for gradient elasticity, viscoplasticity, and damage. Journal of Engineering Mechanics 135, 117–131.

Forest, S., Guéninchault, N., 2013. Inspection of free energy functions in gradient crystal plasticity. Acta Mechanica Sinica 29 (6), 763–772.

Forest, S., Sedláček, R., 2003. Plastic slip distribution in two-phase laminate microstructures: dislocation-based versus generalized-continuum approaches. Philosophical Magazine 83 (2), 245–276.

Franciosi, P., Zaoui, A., 1982. Multislip in F.C.C. crystals; a theoretical approach compared with experimental data. Acta Metall. 30, 1627–1637.

Frank, F. C., 1951. LXXXIII. Crystal dislocations–Elementary concepts and definitions. Philosophical Magazine 42 (331), 809–819.

Fredriksson, P., Gudmundson, P., 2005. Size-dependent yield strength of thin films. International journal of plasticity 21 (9), 1834–1854.

Fredriksson, P., Gudmundson, P., 2007. Modelling of the interface between a thin film and a substrate within a strain gradient plasticity framework. Journal of the Mechanics and Physics of Solids 55 (5), 939–955.

Geers, M., De Borst, R., Brekelmans, W., Peerlings, R., 1998. Strain-based transient-gradient damage model for failure analyses. Computer Methods in Applied Mechanics and Engineering 160 (1), 133–153.

Geers, M. G. D., Brekelmans, W. A. M., Bayley, C. J., 2007. Second-order crystal plasticity: internal stress effects and cyclic loading. Modelling Simul. Mater. Sci. Eng. 15, 133–145.

Gillis, P. P., Gilman, J. J., 1965. Dynamical dislocation theory of crystal plasticity II. Easy glide and strain hardening. Journal of Applied Physics 36, 3380–3386.

Groma, I., Csikor, F., Zaiser, M., 2003. Spatial correlations and higher-order gradient terms in a continuum description of dislocation dynamics. Acta Materialia 51 (5), 1271–1281.

Gruber, P. A., Bohm, J., Onuseit, F., Wanner, A., Spolenak, R., Arzt, E., 2008. Size effects on yield strength and strain hardening for ultra-thin Cu films with and without passivation: A study by synchrotron and bulge test techniques. Acta Materialia 56, 2318–2335.

Gudmundson, P., 2004. A unified treatment of strain gradient plasticity. Journal of the Mechanics and Physics of Solids 52 (6), 1379–1406.

Gurtin, M. E., 2000. On the plasticity of single crystals: free energy, microforces, plastic-strain gradients. Journal of the Mechanics and Physics of Solids 48 (5), 989–1036.

Gurtin, M. E., 2002. A gradient theory of single-crystal viscoplasticity that accounts for geometrically necessary dislocations. Journal of the Mechanics and Physics of Solids 50, 5–32.

Gurtin, M. E., 2004. A gradient theory of small-deformation isotropic plasticity that accounts for the Burgers vector and for dissipation due to plastic spin. Journal of the Mechanics and Physics of Solids 52 (11), 2545–2568.

Gurtin, M. E., 2006. The Burgers vector and the flow of screw and edge dislocations in finite-deformation single-crystal plasticity. Journal of the Mechanics and Physics of Solids 54, 1882–1898.

Gurtin, M. E., 2008a. A finite-deformation, gradient theory of single-crystal plasticity with free energy dependent on densities of geometrically necessary dislocations. International Journal of Plasticity 24 (4), 702–725.

Gurtin, M. E., 2008b. A theory of grain boundaries that accounts automatically for grain misorientation and grain-boundary orientation. Journal of the Mechanics and Physics of Solids 56 (2), 640–662.

Gurtin, M. E., Anand, L., Lele, S. P., 2007. Gradient single-crystal plasticity with free energy dependent on dislocation densities. Journal of the Mechanics and Physics of Solids 55, 1853–1878.

Gurtin, M. E., Needleman, A., 2005. Boundary conditions in small-deformation, single-crystal plasticity that account for the Burgers vector. Journal of the Mechanics and Physics of Solids 53 (1), 1–31.

Hall, E. O., 1951. The deformation and aging of mild steel. Part III: discussion and results. Proceedings of the Physical Society of London 64, 747–753.

Han, C.-S., Ma, A., Roters, F., Raabe, D., 2007. A finite element approach with patch projection for strain gradient plasticity formulations. International Journal of Plasticity 23, 690–710.

Han, W., Reddy, B. D., 2013. Plasticity: mathematical theory and numerical analysis. Vol. 9. Springer, New York.

Havner, K., 1992. Finite plastic deformation of crystalline solids. Cambridge University Press, Cambridge.

Hill, R., 1966. Generalized constitutive relations for incremental deformation of metal crystals by multislip. Journal of the Mechanics and Physics of Solids 14, 95–102.

Hirth, J. P., Lothe, J., 1982. Theory of dislocations. John Wiley & Sons, Canada.

Hochrainer, T., 2006. Evolving systems of curved dislocations: Mathematical foundations of a statistical theory. Ph.D. thesis, Universität Karlsruhe (TH), Aachen.

Hochrainer, T., 2013. Higher order alignment tensors for continuum dislocation dynamics. In: MRS Proceedings. Vol. 1535 (mmm2012–a). Cambridge Univ Press.

Hochrainer, T., Sandfeld, S., Zaiser, M., Gumbsch, P., 2014. Continuum dislocation dynamics: towards a physical theory of crystal plasticity. Journal of the Mechanics and Physics of Solids 63, 167–178.

Hochrainer, T., Zaiser, M., 2005. Fundamentals of a continuum theory of dislocations. In: Proceedings of Science, International conference on Statistical Mechanics of Plasticity and related Instabilities (SMPRI 2005). pp. 1–12.

Hochrainer, T., Zaiser, M., Gumbsch, P., 2007. A three-dimensional continuum theory of dislocations: kinematics and mean field formulation. Philosophical Magazine 87, 1261–1282.

Hochrainer, T., Zaiser, M., Gumbsch, P., 2010. Dislocation transport and line length increase in averaged descriptions of dislocations. arXiv:1010.2884v1.

Hull, D., Bacon, D. J., 1984. Introduction to dislocations. Vol. 257. Pergamon Press, Oxford.

Hurtado, D. E., Ortiz, M., 2012. Surface effects and the size-dependent hardening and strengthening of nickel micropillars. Journal of the Mechanics and Physics of Solids 60 (8), 1432–1446.

Hurtado, D. E., Ortiz, M., 2013. Finite element analysis of geometrically necessary dislocations in crystal plasticity. International Journal for Numerical Methods in Engineering 93 (1), 66–79.

Hutchinson, J., 1976. Bounds and self-consistent estimates for creep of polycrystalline materials. Proceedings of the Royal Society of London A. Mathematical and Physical Sciences 348 (1652), 101–127.

Hutchinson, J. W., 2012. Generalizing J 2 flow theory: Fundamental issues in strain gradient plasticity. Acta Mechanica Sinica 28 (4), 1078–1086.

Kametani, R., Kodera, K., Okumura, D., Ohno, N., 2012. Implicit iterative finite element scheme for a strain gradient crystal plasticity model based on self-energy of geometrically necessary dislocations. Computational Materials Science 53 (1), 53–59.

Klusemann, B., Bargmann, S., Svendsen, B., 2012. Two models for gradient inelasticity based on non-convex energy. Computational Materials Science 64, 96–100.

Kochmann, D. M., Le, K. C., 2008. Dislocation pile-ups in bicrystals within continuum dislocation theory. International Journal of Plasticity 24 (11), 2125–2147.

Kochmann, D. M., Le, K. C., 2009a. A continuum model for initiation and evolution of deformation twinning. Journal of the Mechanics and Physics of Solids 57 (6), 987–1002.

Kochmann, D. M., Le, K. C., 2009b. Plastic deformation of bicrystals within continuum dislocation theory. Mathematics and Mechanics of Solids 14 (6), 540–563.

Kocks, U. F., 1970. The relation between polycrystal deformation and single-crystal deformation. Metall. Trans. 1, 1121–1143.

Kocks, U. F., Mecking, H., 2003. Physics and phenomenology of strain hardening: the FCC case. Progress in Materials Science 48, 171–273.

Koiter, W. T., 1953. Stress-strain relations, uniqueness and variational theorems for elasto-plastic material with a singular yield surface. Quart. Appl. Math. 11, 350–354.

Kondo, K., 1952. On the geometrical and physical foundations of the theory of yielding. In: Proceedings Japan National Congress of Applied Mechanics. Vol. 2. pp. 41–47.

Kröner, E., 1958. Kontinuumstheorie der Versetzungen und Eigenspannungen. Springer, Berlin.

Kröner, E., 1960. Allgemeine Kontinuumstheorie der Versetzungen und Eigenspannungen. Archive for Rational Mechanics and Analysis 4, 273–334.

Kuroda, M., 2013. Higher-order gradient effects in micropillar compression. Acta Materialia 61 (7), 2283–2297.

Kuroda, M., Tvergaard, V., 2008. A finite deformation theory of higher-order gradient crystal plasticity. Journal of the Mechanics and Physics of Solids 56, 2573–2584.

Le, K. C., Sembiring, P., 2008a. Analytical solution of plane constrained shear problem for single crystals within continuum dislocation theory. Archive of Applied Mechanics 78 (8), 587–597.

Le, K. C., Sembiring, P., 2008b. Plane constrained shear of single crystal strip with two active slip systems. Journal of the Mechanics and Physics of Solids 56 (8), 2541–2554.

Liebe, T., 2004. Theory and numerics of higher gradient inelastic material behavior. Ph.D. thesis, Universität Kaiserslautern.

Liebe, T., Menzel, A., Steinmann, P., 2003. Theory and numerics of geometrically non-linear gradient plasticity. International Journal of Engineering Science 41 (13), 1603–1629.

Liebe, T., Steinmann, P., 2001. Theory and numerics of a thermodynamically consistent framework for geometrically linear gradient plasticity. Int. J. Numer. Meth. Engng. 51, 1437–1467.

Liebe, T., Steinmann, P., Benallal, A., 2001. Theoretical and computational aspects of a thermodynamically consistent framework for geometrically linear gradient damage. Computer Methods in Applied Mechanics and Engineering 190 (49), 6555–6576.

Mader, S., 1957. Elektronenmikroskopische Untersuchung der Gleitlinienbildung auf Kupfereinkristallen. Zeitschrift für Physik A Hadrons and Nuclei 149 (1), 73–102.

Mandel, J., 1971. Plasticité classique et viscoplasticité: course held at the Department of Mechanics of Solids, September-October, 1971. Springer.

Mazière, M., Forest, S., 2013. Strain gradient plasticity modeling and finite element simulation of Lüders band formation and propagation. Continuum Mechanics and Thermodynamics, 1–22.

Menzel, A., Steinmann, P., 2000. On the continuum formulation of higher gradient plasticity for single and polycrystals. Journal of the Mechanics and Physics of Solids 48, 1777–1796.

Mesarovic, S. D., 2005. Energy, configurational forces and characteristic lengths associated with the continuum description of geometrically necessary dislocations. International journal of plasticity 21 (10), 1855–1889.

Mesarovic, S. D., 2010. Plasticity of crystals and interfaces: from discrete dislocations to size-dependent continuum theory. Theoretical and Applied Mechanics 37 (4), 289–332.

Mesarovic, S. D., Baskaran, R., Panchenko, A., 2010. Thermodynamic coarsening of dislocation mechanics and the size-dependent continuum crystal plasticity. Journal of the Mechanics and Physics of Solids 58 (3), 311–329.

Miehe, C., 2002. Strain-driven homogenization of inelastic microstructures and composites based on an incremental variational formulation. International Journal for Numerical Methods in Engineering 55 (11), 1285–1322.

Miehe, C., 2011. A multi-field incremental variational framework for gradient-extended standard dissipative solids. Journal of the Mechanics and Physics of Solids 59, 898–923.

Miehe, C., 2014. Variational gradient plasticity at finite strains. Part I: Mixed potentials for the evolution and update problems of gradient-extended dissipative solids. Computer Methods in Applied Mechanics and Engineering 268, 677–703.

Miehe, C., Aldakheel, F., Mauthe, S., 2013. Mixed variational principles and robust finite element implementations of gradient plasticity at small strains. International Journal for Numerical Methods in Engineering 94 (11), 1037–1074.

Miehe, C., Apel, N., Lambrecht, M., 2002. Anisotropic additive plasticity in the logarithmic strain space: modular kinematic formulation and implementation based on incremental minimization principles for standard materials. Computer Methods in Applied Mechanics and Engineering 191 (47), 5383–5425.

Miehe, C., Hofacker, M., Welschinger, F., 2010a. A phase field model for rate-independent crack propagation: Robust algorithmic implementation based on operator splits. Computer Methods in Applied Mechanics and Engineering 199 (45), 2765–2778.

Miehe, C., Mauthe, S., Hildebrand, F., 2014a. Variational gradient plasticity at finite strains. Part III: Local–global updates and regularization techniques in multiplicative plasticity for single crystals. Computer Methods in Applied Mechanics and Engineering 268, 735–762.

Miehe, C., Schröder, J., 2001. A comparative study of stress update algorithms for rate-independent and rate-dependent crystal plasticity. Int. J. Numer. Meth. Engng. 50, 273–298.

Miehe, C., Welschinger, F., Aldakheel, F., 2014b. Variational gradient plasticity at finite strains. Part II: Local–global updates and mixed finite elements for additive plasticity in the logarithmic strain space. Computer Methods in Applied Mechanics and Engineering 268, 704–734.

Miehe, C., Welschinger, F., Hofacker, M., 2010b. Thermodynamically consistent phase-field models of fracture: Variational principles and multi-field FE implementations. International journal for numerical methods in engineering 83 (10), 1273–1311.

Mindlin, R., 1964. Micro-structure in linear elasticity. Archive for Rational Mechanics and Analysis 16 (1), 51–78.

Needleman, A., Asaro, R., Lemonds, J., Peirce, D., 1985. Finite element analysis of crystalline solids. Computer Methods in Applied Mechanics and Engineering 52 (1), 689–708.

Nielsen, K. L., Niordson, C., 2013. A 2D finite element implementation of the Fleck–Willis strain-gradient flow theory. European Journal of Mechanics-A/Solids 41, 134–142.

Nix, W. D., Gao, H., 1998. Indentation size effects in crystalline materials: a law for strain gradient plasticity. Journal of the Mechanics and Physics of Solids 46 (3), 411–425.

Nye, J. F., 1953. Some geometrical relations in dislocated crystals. Acta Metallurgica 1, 153–162.

Ohno, N., Okumura, D., 2007. Higher-order stress and grain size effects due to self energy of geometrically necessary dislocations. Journal of the Mechanics and Physics of Solids 55, 1879–1898.

Orowan, E., 1934. Zur Kristallplastizität III. Zeitschrift für Physik 89, 634–659.

Ortiz, M., Repetto, E., 1999. Nonconvex energy minimization and dislocation structures in ductile single crystals. Journal of the Mechanics and Physics of Solids 47 (2), 397–462.

Ortiz, M., Stainier, L., 1999. The variational formulation of viscoplastic constitutive updates. Comput. Methods Appl. Mech. Engrg. 171, 419–444.

Peerlings, R., Poh, L., Geers, M., 2012. An implicit gradient plasticity–damage theory for predicting size effects in hardening and softening. Engineering Fracture Mechanics 95, 2–12.

Perzyna, P., 1971. Thermodynamic theory of viscoplasticity. Advances in Applied Mechanics 11, 313–354.

Petch, N. J., 1953. The cleavage strength of polycrystals. Journal of the Iron and Steel Institute 174, 25–28.

Poh, L., Peerlings, R., Geers, M., Swaddiwudhipong, S., 2011. An implicit tensorial gradient plasticity model–formulation and comparison with a scalar gradient model. International Journal of Solids and Structures 48 (18), 2595–2604.

Polanyi, M., 1934. Über eine Art Gitterstörung, die einen Kristall plastisch machen könnte. Zeitschrift für Physik 89 (9-10), 660–664.

Portevin, A., Le Chatelier, F., 1923. Sur un phénomène observé lors de l'essai de traction d'alliages en cours de transformation. Comptes Rendus de l'Académie des Sciences Paris 176, 507–510.

Reddy, B. D., Wieners, C., Wohlmuth, B., 2012. Finite element analysis and algorithms for single-crystal strain-gradient plasticity. International Journal for Numerical Methods in Engineering 90 (6), 784–804.

Rice, J. R., 1971. Inelastic constitutive relations for solids: an internal-variable theory and its application to metal plasticity. Journal of the Mechanics and Physics of Solids 19, 433–455.

Sandfeld, S., 2010. The evolution of dislocation density in a higher-order continuum theory of dislocation plasticity. Ph.D. thesis, The University of Edinburgh.

Sandfeld, S., Hochrainer, T., Zaiser, M., Gumbsch, P., 2010. Numerical implementation of a 3D continuum theory of dislocation dynamics and application to micro-bending. Philosophical Magazine 90, 3697–3728.

Sandfeld, S., Hochrainer, T., Zaiser, M., Gumbsch, P., 2011. Continuum modeling of dislocation plasticity: Theory, numerical implementation, and validation by discrete dislocation simulations. J. Mater. Res. 26, 623–632.

Schmid, E., Boas, W., 1935. Kristallplastizität: mit besonderer Berücksichtigung der Metalle. Vol. 17. Springer, Berlin.

Schob, O., 1972. Zur Kristallographie der Gleitverformung von Wolfram-Einkristallen im Zugversuch. Monatshefte für Chemie/Chemical Monthly 103 (1), 255–269.

Shu, J., Fleck, N., Van der Giessen, E., Needleman, A., 2001. Boundary layers in constrained plastic flow: comparison of nonlocal and discrete dislocation plasticity. Journal of the Mechanics and Physics of Solids 49 (6), 1361–1395.

Simo, J. C., Hughes, T., 1998. Computational Inelasticity. Springer, New York.

Steinmann, P., 1996. Views on multiplicative elastoplasticity and the continuum theory of dislocations. Int. J. Eng. Sci. 34, 1717–1735.

Steinmann, P., Stein, E., 1996. On the numerical treatment and analysis of finite deformation ductile single crystal plasticity. Computer methods in applied mechanics and engineering 129 (3), 235–254.

Stölken, J. S., Evans, A. G., 1998. A microbend test method for measuring the plasticity length scale. Acta mater. 46, 5109–5115.

Svendsen, B., Bargmann, S., 2010. On the continuum thermodynamic rate variational formulation of models for extended crystal plasticity at large deformation. Journal of the Mechanics and Physics of Solids 58 (9), 1253–1271.

Taylor, G. I., 1934. The mechanism of plastic deformation of crystals. Part I. Theoretical. Proceedings of the Royal Society of London A 145, 362–387.

Taylor, G. I., 1938. Plastic strain in metals. Journal of the Institute of Metals 62, 307–324.

Teodosiu, C., 1970. A dynamic theory of dislocations and its applications to the theory of the elastic-plastic continuum. In: Fundamental Aspects of Dislocation Theory. National Bureau of Standards (U.S.), pp. 837–876, special Publication, Vol., II.

Volterra, V., 1907. Sur l'équilibre des corps élastiques multiplement connexes. In: Annales scientifiques de l'Ecole Normale supérieure. Vol. 24. Société mathématique de France, pp. 401–517.

Voyiadjis, G., Deliktas, B., 2009. Formulation of strain gradient plasticity with interface energy in a consistent thermodynamic framework. International Journal of Plasticity 25 (10), 1997–2024.

Wieners, C., Wohlmuth, B., 2011. A primal-dual finite element approximation for a nonlocal model in plasticity. SIAM J. Numer. Anal. 49, 692–710.

Wulfinghoff, S., Bayerschen, E., Böhlke, T., 2013a. A gradient plasticity grain boundary yield theory. International Journal of Plasticity 51, 33–46.

Wulfinghoff, S., Bayerschen, E., Böhlke, T., 2013b. Micromechanical simulation of the Hall-Petch effect with a crystal gradient theory including a grain boundary yield criterion. PAMM 13 (1), 15–18.

Wulfinghoff, S., Bayerschen, E., Böhlke, T., 2013c. Modeling the Hall-Petch effect with a gradient crystal plasticity theory including a grain boundary yield criterion. In: Proceedings of the XII International Conference on Computational Plasticity, COMPLAS XII.

Wulfinghoff, S., Böhlke, T., 2010. Gradient plasticity for single crystals. PAMM, Proc. Appl. Math. Mech. 10, 351–352.

Wulfinghoff, S., Böhlke, T., 2012a. Equivalent plastic strain gradient enhancement of single crystal plasticity: theory and numerics. Proceedings of the Royal Society A: Mathematical, Physical and Engineering Science 468 (2145), 2682–2703.

Wulfinghoff, S., Böhlke, T., 2012b. Three-dimensional continuum dislocation microplasticity FE-simulation. In: Eberhardsteiner, J., Böhm, H., Rammerstorfer, F. (Eds.), CD-ROM Proceedings of the 6th European Congress on Computational Methods in Applied Sciences and Engineering (ECCOMAS 2012). Vienna University of Technology, Austria, ISBN: 978-3-9502481-9-7.

Wulfinghoff, S., Böhlke, T., 2013. Equivalent plastic strain gradient crystal plasticity – enhanced power law subroutine. GAMM-Mitteilungen 36 (2), 134–148.

Wulfinghoff, S., Forest, S., Böhlke, T., 2014. Logarithmic and rank-one defect energies in gradient crystal plasticity – analytical and numerical solutions for laminate microstructures (submitted).

Wulfinghoff, S., Glavas, V., Böhlke, T., 2011. Dislocation transport in single crystals and dislocation-based micromechanical hardening. PAMM, Proc. Appl. Math. Mech. 11, 449–450.

Xiang, Y., Vlassak, J., 2006. Bauschinger and size effects in thin-film plasticity. Acta Materialia 54, 5449–5460.

Yalcinkaya, T., Brekelmans, W. A. M., Geers, M. G. D., 2008. BCC single crystal plasticity modeling and its experimental identification. Modelling Simul. Mater. Sci. Eng.1 16, 085007 (16pp).

Yang, B., Motz, C., Rester, M., Dehm, G., 2012. Yield stress influenced by the ratio of wire diameter to grain size–a competition between the effects of specimen microstructure and dimension in micro-sized polycrystalline copper wires. Philosophical Magazine 92 (25-27), 3243–3256.

Yassar, R. S., Mesarovic, S. D., Field, D. P., 2007. Micromechanics of hardening of elastic–plastic crystals with elastic inclusions: I–Dilute concentration. International Journal of Plasticity 23 (10), 1901–1917.

Yefimov, S., Groma, I., Van der Giessen, E., 2004. A comparison of a statistical-mechanics based plasticity model with discrete dislocation plasticity calculations. J. Mech. Phys. Solids 52, 279–300.

Zaiser, M., Nikitas, N., Hochrainer, T., Aifantis, E. C., 2007. Modelling size effects using 3D density-based dislocation dynamics. Philosophical Magazine 87, 1283–1306.

Schriftenreihe Kontinuumsmechanik im Maschinenbau
Karlsruher Institut für Technologie (KIT)
(ISSN 2192-693X)

Herausgeber: Prof. Dr.-Ing. Thomas Böhlke

Die Bände sind unter www.ksp.kit.edu als PDF frei verfügbar oder als Druckausgabe bestellbar.

Band 1 Felix Fritzen
Microstructural modeling and computational homogenization of the physically linear and nonlinear constitutive behavior of micro-heterogeneous materials. 2011
ISBN 978-3-86644-699-1

Band 2 Rumena Tsotsova
Texturbasierte Modellierung anisotroper Fließpotentiale. 2012
ISBN 978-3-86644-764-6

Band 3 Johannes Wippler
Micromechanical finite element simulations of crack propagation in silicon nitride. 2012
ISBN 978-3-86644-818-6

Band 4 Katja Jöchen
Homogenization of the linear and non-linear mechanical behavior of polycrystals. 2013
ISBN 978-3-86644-971-8

Band 5 Stephan Wulfinghoff
Numerically Efficient Gradient Crystal Plasticity with a Grain Boundary Yield Criterion and Dislocation-based Work-Hardening. 2014
ISBN 978-3-7315-0245-6